SILENT SEAS

SILENT SEAS

The Fish Race to the Bottom

ISABELLA LÖVIN

Paragon Publishing 2012

Published by Isabella Lövin

Publishing partner: Paragon Publishing, Rothersthorpe

First published 2012

ISBN 978-1-908341-53-2

Originally published as *Tyst hav – Jakten på den sista matfisken*, in Swedish in 2007 by Ordfront förlag

Translation into English courtesy Baltic Sea 2020 Foundation

Cover photograph *Superstock*

Cover design *Åsa Lundqvist*

Book design, layout and production management by Into Print

www.intoprint.net

+44 (0)1604 832149

Printed and bound in UK and USA by Lightning Source

Chain of Custody certification from: The Forest Stewardship Council, Programme for the Endorsement of Forest Certification and The Sustainable Forestry Initiative

To Ester and Gunnar

"So the present is linked with future and past,
and each living thing with all
that surrounds it."
RACHEL CARSON, *The Edge of the Sea*

CONTENTS

FOREWORD

December 2011

No one asked me to write *Silent Seas*. No editor believed in the idea when I presented it to them, and most of them didn't believe in it even after I had the manuscript ready. All except one – and so finally the book was published in Sweden in the autumn of 2007, with a first very tiny edition.

What happened next was surprising even to me – who of course was convinced that this topic should be of enormous interest to everybody even the slightest bit interested in environmental or political issues. But I didn't foresee the overwhelming attention this book would attract of every possible organisation in Sweden that had anything to do with food, seas, water, environment, journalism, politics, animal welfare or fisheries of course. It was awarded no less than 14 different prizes; it was praised and debated continuously in the media for months, and I was interviewed almost every day for around two years after the book was released, and the whole debate climate on fisheries issues has radically changed in the country.

Four years have now elapsed since the book was published in Sweden, and these years have been the most intense of my life so far. I was asked by the Swedish Green Party to be their candidate in the elections for the European Parliament in 2009 and, after much hesitation, I accepted. I was indeed elected, and am now strangely enough a member of the Fisheries Committee in the European Parliament, the same Committee I describe with such fascination in this book when I visited parliament in 2004.

I choose to publish this book in English so many years
published in Swedish? Is it still of any interest at all? Are
blems stated in the book still valid?

Depressingly enough I have to answer yes to the latter. And to the first – why hasn't this book appeared in English until now? Well – it seems that English and US publishers have a lot in common with their Swedish counterparts: in a difficult economic situation, most publishers are looking for safe returns on their investment, and not even an award-winning book like this seems safe enough, since there are already other books on the market dealing with overfishing.

But the book is finally here, partly thanks to the Baltic Sea 2020 Foundation, and although I am aware that a few chapters are very much Sweden-focused, I hope that they still can be read as case studies, and that they reflect a reality of close relationships between the fishery industry and decision makers that can be translated to many other countries around the world. All the facts have been reviewed and I have added some parentheses with updated comments from 2011.

It is my sincere hope that this book will contribute to a wider, public European debate on the future of the Common Fisheries Policy – knowing that the window of opportunity to influence what is going on behind the scenes is closing in this very moment, and that it's closing *fast*.

A new EU Common Fisheries Policy should be in place the 1st January, 2013 – if everything goes according to timetable – and the proposal for a new policy was already presented by the EU Commission in the summer of 2011. Europeans now have less than one year remaining to influence politicians to be bold enough to take the right, radical decisions so that we can turn this destructive, vicious circle that is the EU fisheries policy into a virtuous and sustainable one.

The Commissioner for Fisheries, Maria Damanaki (in this book only mentioned in the foreword, since she took office only in 2009) has pointed out that without a reform, only 8 out of 136 European fish stocks will be healthy in 2022. Why did she point out 2022 of all the possible years? Well, because that will be the next opportunity for European decision makers to make it right if we don't succeed this time. The reformed Common Fisheries Policy (CFP) will stay in place for ten long years. The last time it was reformed was in 2002 and since then stocks have continued to decline and catches gone down, as well as profits – while tax payers' money has been generously spent on "restructuring" the EU fleet – with the result that fishing capacity has increased by three per cent each year! If this trend continues, by 2022 we will end up having an enormous number of boats and fishermen, but virtually no fish left in the seas. And this is not a worst-case-scenario; this is *the* scenario.

Since the original publication of this book in 2007, there has been only one positive change in the status of stocks in European seas, but there have been several negative ones. The eastern cod stock in the Baltic Sea actually seems to be recovering now, to a large extent due to favourable natural conditions, but also in fact due to a distinct political will to do something about both legal and illegal overfishing. I am glad to say that this book probably contributed to the strong positions taken by the Swedish government in this respect, something that definitely gives me hope that things actually can change – if only there are strong public opinions in the different European countries.

Less encouraging is the status of the European eel; even three years after the EU eel recovery plan was put in place, the stock continues to decrease – according to the last assessment from the International Council of the Exploration of the Seas (ICES), the stock is now at an historic low; and even if all fisheries were to be stopped, there is no guarantee that it would ever recover. The status of cod in the North Sea is also very worrying, and new evidence shows that a number of North Sea stocks has actually been wiped out. Spiny dogfish has finally been protected, which is good news of course, but only after a decline of more than 95 per cent – which is bad news.

The depressing list can be made very long, but it is more important now to point out those intrinsically destructive mechanisms of European overfishing that need to be corrected. The major one is called Council negotiations. Commissioner Maria Damanaki has pointed out this very destructive way of managing shared fish stocks by yearly negotiations in the Council of Ministers, calling it "horse-trading" (= "if you let me overfish, I'll let you overfish", or even: "if you let me overfish, I'll give you higher milk quotas"). Her very strong ambition when she came into office was to put an end to this praxis of neglecting scientific advice, and she came out very strongly announcing that the Commission demanded that the Council of Ministers accept the quotas proposed by the Commission, based on scientific advice. But looking at the result of the last December Council meeting (2011), where yearly quotas of fish catches were decided, the outcome was, depressingly enough, very similar to the previous ones. In 79 per cent of the cases the ministers decided on quotas higher than the Commission had proposed, sometimes almost incomprehensibly much higher – as in the case of boarfish where the Commission proposed a 15 per cent reduction, and the Council instead decided on a 148 per cent increase!

When I was working on this book I kept asking myself with what right policy makers ignore the best available scientific advice and take the liberty of selling out our common property, the seas, to such a small interest group as the fishermen? I looked at the figures, trying to find the economic explanation for this strange behaviour. What I found puzzled me even more, and to be honest, I wasn't completely sure even when the manuscript went to the printers that what I had found wasn't a bit too biassed or that I simply hadn't just forgotten about some important economic aspect of the whole business. But all the figures I dug up for the chapter called "Subsidies" turned out to be true, and the Swedish National Auditors actually did a big revision of the Swedish fisheries sector in 2008, which confirmed the conclusion of this book: that the fishing industry is actually costing taxpayers almost as much as the total added value of the whole sector. The Swedish National Auditor, Eva Lindström, was surprised that most of the fishermen could even cover their operation costs, seeing how little they declared as incomes. And she also singled out the tax break on fuel as a perverse incentive for fishermen to go on fishing when stocks are depleted, making them even less sensitive to increasing fuel prices than other sectors.

The fishing sector in fact has alarmingly high CO_2 emissions compared to other sectors, put in relation to its contribution to GDP, which was also a point made by the National Auditors. On average in Sweden all the sectors emit 280 kilos of CO_2 per million euros earned, while the fishing sector emits 4,610 kilos of CO_2 per million euros! The sector, by the way, leads in another league in Sweden, being the one with the most "black" money circulated; when the Swedish Tax Authority made a top ten list, the fishing sector was way ahead of farmers, restaurant owners, hairdressers, carpenters and taxi drivers in hiding money from taxation.

After two years in the European parliament, I have learnt nothing that indicates that Sweden and its fisheries sector is the black sheep of Europe. On the contrary, control of the fisheries sector is weak everywhere, as well as the profitability of the sector. According to the EU Commission 35 per cent of the businesses are operating at a loss and overall, in 2009, the entire EU fleet was making almost no profit at all, not even if subsidies were counted. In the major fishing power of Europe, Spain, each person employed in fishing was subsidised by 27,000 euros. Even with the subsidies the Spanish fleet was losing 250 million euros after deducting operating and capital costs from income. This is money that could have been spent on far better ways of creating sustainable jobs than just keeping too many fishermen busy chasing too few fish. According to the Commission, each year the EU loses 2.7 billion euros because of the overfished status of our seas!

But this book is not only about money, far from it. Mostly it is about far greater values that are being gambled away by irresponsible politicians. One such value is global justice. What happens to us in Europe when we deplete our own fish stocks? Do we eat less fish? The answer is no. Absolutely not! Although EU catches have gone down by almost 40 per cent since 1995, we are eating more and more sushi, fish fingers and bluefin tuna than ever – consumption per capita is on a sharp increase. So – where does all this fish come from? The answer is that we import more and more. Now over 60 per cent of all fish consumed in the EU is imported, and a large share of that comes from developing countries. So, by irresponsibly emptying our own nearby waters, we are now causing an increasing exploitation of the waters off Western Africa and in the Indian and Pacific Oceans. Overfishing in Europe is depriving poor people around the world of their much needed marine protein. The big trawlers moving around the world often even come from Europe, in other cases it is just the buyers that do. On the local fish markets in Bissau or Dakar there are no big fish anymore, only small, less valuable "Africa-fish". All the big species are immediately flown to Europe where people can afford to pay more for them; the fishmongers call them "Europe-fish".

The justice and food security issues are aspects of overfishing that dawned on me fairly late in the process of writing this book, but they are now a large part of what keeps me going. Another extremely serious aspect of overfishing on a global scale, which also didn't become clear to me until recently, is the connection between fish and climate change. In December 2009 the UN-report *Blue Carbon* was released, and it suddenly filled in some of the crucial missing pieces in the complex picture that is the earth's climate cycle. The oceans still haven't been entirely included in the calculations that formed the basis of climate negotiations in Copenhagen, Cancun or Durban. But the *Blue Carbon* report shows that the oceans are as important as the entire terrestrial world when it comes to oxygen production and CO_2 sequestration. And the oceans are losing these capacities at an alarming speed. Due mainly to coastal water degradation, including overfishing, the world's oceans are losing more of their capacity to absorb CO_2 each year. Healing the oceans by reducing pollution, restoring mangroves and letting the big fish grow back would be a good investment in the future of this planet; at least as important as conserving the rainforests.

Fish is so much more than just food. All the marine creatures, from zooplankton to whales – with fish, dolphins and sharks in-between – are actually crucial agents in an ecological system where each species has its own important role. If we remove some of these species, or deplete them to just fractions of what they used to be, we of course also change the functioning and decrease the resilience of that system. And since the system we are talking about has evolved during a time span of some 4.6 billion years, and the size of it is more than 98 per cent of the entire planet's biosphere – taking into account that the living space in the sea is much vaster than the more two-dimensional earth surface – I say we are toying with huge values, gambling with functions of this planet that may not be repairable – and for what?

The EU is one of the three biggest fishing powers in the world, after China and Peru. We are, in addition to this, the largest fish consumption market in the world: we have got lots of power – and responsibilities – both through our fisheries and our market policies.

Now this superpower of fisheries is finally reforming its policies. The window of opportunity is open for a few more months. So, let's not waste any more time. Let's make winds of change finally blow through that window – and always remember: the sea is our common heritage; the sea belongs to us all.

THE EEL

"It's your fault," says the man from Glass Eels Ltd. unexpectedly. Smiling uncomfortably, he gesticulates towards me, sitting in the front row of the auditorium at the Observatory Museum in Stockholm. It's October 2005 and I am probably the only journalist attending this packed symposium on the future of the European eel. Hosting the event is Axel Wenblad, recently appointed Director-General of the Swedish Board of Fisheries; a newcomer in the world of fisheries, coming from the business sector and with a background as a biologist. Shocked, like me, to learn of the dramatic decline in European eel populations in the last twenty years, Wenblad now has summoned eel fishery representatives to discuss the problem, and to look at PowerPoint slides with curves that all point in one direction – downwards. The figures shown are not even local projections or snapshots of worst scenarios but absolute figures, showing the entire eel population: a decline of more than ninety-nine per cent.

That's right. *More* than ninety-nine per cent.

"It's your fault, Isabella," repeats the man from UK-based Glass Eels Ltd. stiffly. The white plastic tag on my jacket obviously has caught his eye, and because of it he is able to use my name. He is in his sixties, dressed in a navy blue suit and tie – a fact that makes him stand out from this particular crowd. All the leading eel experts in Sweden and Europe are here – scientists from the UK, the Netherlands and France, experts from WWF and the European Union, fishermen, politicians, anglers and hydropower industry representatives, geneticists and bureaucrats. What these people don't know about eels is not worth knowing; all the figures, diagrams, percentages, costs, trends, risks and historical data are at their fingertips. A striking number of them are wearing khaki jackets, woollen jumpers and thick rubber-soled shoes; everyone except people representing the EU and the fishing industry: they are all wearing suits.

The man from Glass Eels Ltd. peers over his glasses at the seventy or so people in the audience and now points at a man sitting next to me.

"It's your fault too."

Clearing his throat, he moves his finger again, points at the next person and repeats the same words. It all sounds a bit wooden, flat, as if he's practised the speech in front of the bathroom mirror but failed to inject life into it even then.

"And yours," he continues, gesturing hesitantly into the middle of the auditorium. "It's everyone's fault. Yours, mine, Isabella's, our parents – we're all to blame for the modernisation of society that's brought us to this point."

"Do we want a modern society?" he asks rhetorically, nodding and fumbling with the manuscript he's just recited verbatim. He looks so uncomfortable that not even his supporters – the eel fishermen – can rouse themselves to murmur approval.

This unfortunate representative of the glass eel industry (a misleading term because glass eels cannot be produced "industrially" at all, but more of that later) would have needed superior speaking skills and aplomb to challenge the data already presented by the French, Dutch and Swedish scientists. But his heroic presentation "Is there a future for the eel fishery? A view from the industry", is certainly not lacking in detail.

We watch slide after slide of hydropower turbines, mills and drained fields where the eels once used to occur, and listen to him insisting and insisting again and again that *this* is where the problem lies. If we want arable land, electricity, food on our tables and hot water in our taps – then we will have to sacrifice the eel. His own business is just a minor part of the problem: we all must share the blame.

The only problem with this description of events is that it doesn't fit the data we have just heard from the scientists in the room. The industrialisation that brought us water turbines and the agricultural intensification that drained our meadows has been going on for centuries. Yet not until the early 1980s did we see the drastic decline in glass eel numbers in Europe – a decrease that has accelerated in step with the hundreds of millions of euros Brussels has pumped in to modernise the European Union's fishing fleet. Throughout the twentieth century the glass eel population waxed and waned as animal populations usually do. But in the last twenty-five years the curve has headed in one direction only – downwards. In 2003, the influx of glass eels from the Sargasso Sea was less than one per cent of what it was in 1980.

I've been fascinated by fisheries ever since a press release from the Swedish Board of Fisheries landed on my desk in October 2003. I read it once, then read it again, and finally realised I was actually seeing straight. The press release said that the number of eels entering European waters from the Sargasso Sea had plunged by ninety-nine per cent in little more than twenty years. But astonishingly enough, the Board of Fisheries claimed that a moratorium on eel fishing was not justifiable.

Why?

Because it would be unfair to Swedish eel fishermen.

Proposed measures by the authorities?

None.

Still with my head in the normal world, where dire warnings of this sort are taken seriously, I assumed there must be something I didn't quite understand. So I decided to do some research. I got hold of some books about the sea and started reading up about eels – and became completely fascinated. I discovered that eels are one of the oldest animals on the planet – a unique, mystical creature, for a long time not even considered to be a fish because it behaved so mysteriously and could cross land. (Linnaeus gave it the Latin name *Anguilla*, meaning "small snake".) The eel's reproductive cycle remained shrouded in mystery until the end of the 19th century since no one had ever seen an eel fry, adult eels mating or even a sexually mature eel. Aristotle believed that eels emerged from the bowels of the earth, while Pliny suggested they emanated from skin shed by adult eels. A common belief was that eels came from horsehair which had landed in water.

Not until the late 19th century did scientists start to unravel the mystery, after trapping a small, transparent, leaf-shaped monster with strange jaws in the middle of the Atlantic Ocean. The creature – an eel larva – was put in a tank, where it metamorphosed into a glass eel and then into a yellow eel. There and then, one of the strangest biological life cycles on the planet began to unlock. The species proved to be a living fossil that challenged all understanding of the boundaries of biology and evolution. It held the key to a new theory on the geological history of the planet: a prehistoric creature that painted our own short life on Earth in a new light.

We now know that eels mate and spawn in the Sargasso Sea, east of the West Indies and more than 6,000 kilometres away from Sweden. Both American and European eels spawn there. Virtually identical in appearance, they were until recently regarded as the same species, the only difference being that some larvae were carried on the Gulf Stream to Europe (*Anguilla anguilla*) while others drifted on to American shores (*Anguilla rostrata*).

It takes an eel larva at least three years to reach European coasts. Prior to arrival, the leaf-like fry metamorphose into transparent glass eels. (Interestingly, their American cousins undergo the same change a couple of years earlier, before reaching American shores.) Ocean currents carry some glass eels through the Straits of Gibraltar into the Mediterranean, from where they reach the Rhône, the Po or other rivers. Others continue north, towards Scandinavia. Males remain in southern Sweden, while the significantly larger females migrate farther north. It is thought that eels can become male or female depending on where they stay, but that's another story.

Most eels search for fresh water and some even find their way into inland waters by crossing wet grasslands under cover of night. On reaching their destination they remain in a lake and feed on frogs, crustaceans, fish larvae and similar organisms (they are *omnivorous*, eat everything, will not turn their noses up at a water vole or even a baby bird!) with the objective of becoming as fat as possible. At this stage they acquire a yellowish greyish skin and are called yellow eels. Scientists have yet to discover how long eels stay in their freshwater habitat. Periods of seven, ten or even thirty years have been proposed. To complicate matters further, many yellow eels stay in the sea at this phase and do not enter fresh water at all. We do not know how long eels live, but eels in captivity can reach a grand old age. The oldest specimen recorded lived in a tank for 84 years. Another is thought to have survived in a well in southern Sweden for one hundred and fifty years and indeed, when a reporter from Swedish television checked in 2009, it was still there!

Anyway, and here the real miracle begins, eventually the eel metamorphoses again. Its head narrows, its mouth gets smaller, its eyes widen and its back turns black and the belly pale with a silver sheen. And then one autumn night, a few days before full moon, the "silver eel" (as it is now called), gets possessed by its migratory urge and starts to return the same way it once came. Wriggling and swimming to reach the sea again, joining other eels on its way, getting out of the Baltic Sea through the Öresund Strait between Sweden and Denmark, across the North Sea, past the British Isles, out into the Atlantic, across the Atlantic's mid-ocean ridge, and finally several hundred more miles across the deep ocean to its birthplace in the Sargasso Sea. An incredible performance through an endless blue universe with no landmarks or signposts, perhaps navigating with the help of a magnetic crystal in its head, or using its sense of smell to find its way. During this migration the eel does not eat, relying solely on energy-rich fat stored during its time as a yellow eel. On reaching the Sargasso Sea it completes its final transformation, gaining a bronze pigment and mating at an unknown depth (the ocean is around six thousand metres deep here). Then it dies, again in an unknown place. No human being has ever observed an eel mating.

The likely explanation for this extraordinary journey is perhaps even more astonishing than the journey itself. The eel is one of the oldest animals on the planet, dating back 200 million years to a time when the supercontinent Pangaea began to break up. The spawning grounds of these primordial eels are believed to have been the small stretch of water separating what later became Europe and America. As the gap between the continental plates gradually widened over the course of millennia, the eel followed, and every year, every generation, the distance got just a little bit longer. Some eels went east and some went west, genetically programmed to search for their feeding grounds in fresh waters, and all returning faithfully to spawn in their forebears' mating grounds, no matter how long the journey.

With this in mind, it is not surprising to read that the Sargasso Sea, the presumed spawning ground for eels, is in the centre of the North Atlantic current system, adjacent to the Bermuda Triangle. Thus, eels have their birthplace in the navel of the continents, at the dawn of time. They pre-date dinosaurs and have survived ice ages. They have seen lobe-finned fish ascend from sea to land and *Homo sapiens'* appearance in Europe just 40,000 years ago. And now, in the space of just a few short years, they are facing extinction at human hands.

The International Council for the Exploration of the Sea (ICES), a government-financed group of marine biologists from all Atlantic coastal nations, highlighted the plight of the eel back in 2003, stressing that European eels are especially vulnerable as they all belong to the same population. "The first priority is to get the message across to fishermen, managers, and politicians that *the eel, the most widespread and most exploited fish stock in Europe, is dangerously close to collapse,*" ICES wrote.

The scientists noted that eels were fished more intensively at all stages of their life than any other fish species in Europe. Glass eels, yellow eels and silver eels provide an important income for some 25,000 European fishermen. They are fried and cooked with garlic, eaten raw or grilled, served as tapas in Portugal and Spain and as *kabayaki* in Japan. Huge numbers of glass eels are exported live to China. In 2005, companies like Glass Eels Ltd. earned a lucrative rate of 870 euros per kilo from this trade. In Sweden, yellow eels and silver eels are caught on their way back to the Sargasso Sea to be served as a classic *smorgasbord* dish.

(Despite the well known critical state of the eel population, it was not until 2007 that the European Union introduced any plan on restricting eel fisheries. In May 2007, Sweden introduced a moratorium on eel fishing for every citizen – apart from professional fishermen.)

In Sweden, unregulated fishing and increasingly effective fishing methods saw eel catches plummet from more than two and a half thousand tonnes in the 1960s to around six hundred tonnes in 2006. The figures speak volumes, and they are even more alarming considering that the decrease in catches does not reflect the size of the current population, only the *recruitment* of young eels, i.e. the inflow of glass eels (including restocking!) that were still larvae ten, twenty or thirty years ago.

The impact of today's record-low inflows of glass eels has yet to be seen.

The Swedish Board of Fisheries press release that landed on my desk continued thus: "As an emergency initiative, the European Commission recommends a ban on all fishing of silver eels. Such a ban would hit Swedish fisheries very hard. The Board's view is that a balanced approach needs to be taken across all types of eel fisheries. There is not much to be gained from increasing the number of spawning eels in the Sargasso Sea if the glass eels are immediately captured for direct consumption or for export to Chinese fish farms. Nor does it make any difference if an eel is caught this year or one year earlier."

The Board highlighted the restocking programme that Sweden has been operating since the 1970s and which is designed to maintain eel fisheries, noting that the Baltic Sea is an optimal environment for the reintroduction of glass eels.

Based on my new-found knowledge about this extraordinary fish, I got a bit suspicious about the issue of restocking and decided to consult eel experts on the subject. I found that restocking can be good for eel *fisheries* in the short term, as the eels come from local glass eel surpluses in Europe that would not survive in that habitat if they were not moved. But it is highly uncertain if restocking helps the eel *population* in the long run. This is because it has not been established that the glass eels caught, for instance, in the Bristol Channel or the Bay of Biscay, and then released in the Baltic Sea, could find their way back to the Sargasso Sea when it is time for them to spawn. Scientists believe they acquire their migratory compass en route to Europe, making it impossible for them to return to the Sargasso if they are transported by air or truck.

Studies have since confirmed these suspicions. Not one single eel of those marked and released in a lake on the Baltic island of Gotland has been found as a silver eel migrating through the Öresund Strait en route to the Atlantic. Many marked silver eels though have been found swimming southbound until colliding with the German and Polish coasts, apparently completely disoriented. Another study found that some eels could, theoretically, find their way out of the Baltic, but a similar study a couple of years later found that virtually none do so in reality.

Eel farming trials in Europe have been an abject failure. It has turned out that it is extremely difficult to get eels to produce roe and milt (sperm) with artificial hormones. Isolated spawning successes have been recorded, but the larvae then died. The Japanese are further ahead, but eel farming is hugely complex and expensive and not seen as commercially viable. If this extremely complicated creature is going to reproduce it seems it needs to spawn in an undisturbed environment, at a certain depth, at unknown underwater pressure, at a certain salinity, temperature and nutrient level, under a full moon, and in the Sargasso Sea.

Otherwise it simply is not in the mood.

The peculiar issue of the full moon comes up at the eel seminar at the Observatory Museum in Stockholm. The European Commission has a plan. But as Kenneth Patterson from the EU Fisheries Directorate explains the initiative, entitled "EU Management – Current Situation and Future Action", the questions mount. The plan, which took years to develop amid lobbying by Member States and the likes of Glass Eels Ltd. and Scandinavian Silver Eel AB, is so complex and diffuse that now in practice every member state can decide its own approach. Its mysterious formula is that forty per cent of all the silver eels that would have been present in an aquatic environment were it not for human activity, should have free passage to the Sargasso Sea (meaning not fished or killed) to spawn. And those countries that fail to present an action plan by July 1, 2007 to meet this requirement will be subject to a directive banning all eel fishing throughout Europe for the first fifteen days of every month.

So complex is the plan that the Q&A session starts in a long silence as people try to understand the concept. The most obvious problem is how to define and calculate what the plan means by "the level of adult eels which, in the absence of fishing and other effects of human activities, would escape to the sea to spawn". How many eels would there be if there were no humans? If no hydropower plants, marinas, industrial effluent or fishermen existed?

The equation is obviously impossible to solve. The eel is one of the planet's great survivors, capable of swimming six thousand kilometres without food, so full of energy and muscles that it jumps and wriggles long after it has been killed – how many of them would there be if we did not exist? If humans were not part of the world's ecosystems at all, and all other fish and animals also existed without human interference?

The EU Commission, in the shape of Kenneth Patterson, a soberly dressed bureaucrat with a dark beard and glasses, is seemingly unmoved by the philosophical dimensions of this question. He has full confidence, he says, that scientists in each EU member state can perform this calculation in a satisfactory manner and work out what forty per cent of an estimated virgin eel population is.

One member of the audience (a scientist, judging by his attire) is sceptical though. Surely, he says, the figure is too difficult to calculate and a ban on eel fishing will therefore be required in the first two weeks of every month in all the EU member states.

"But have you thought about the moon?" he asks.

Patterson, who has doubtless heard it all during his years spent in EU fishery corridors, stiffens in surprise.

"The moon?"

"Yes," the scientist says. "I've checked my calendar and if the Commission's plan had been effective this year it would have had virtually no impact on the fishery, because full moons are at the end of the month. In 2006, on the other hand, the full moon falls early in the month so there wouldn't be any eels caught at all."

Patterson looks incredulous.

"Well. Everyone knows that eel fishing is linked to the lunar cycle," the scientist continues, and no one in the expert audience contradicts him.

Patterson pulls himself together quickly, claiming that conditions vary between geographic areas, but the French and Dutch delegates do not agree.

The moon, apparently, has been forgotten.

Nobody seems surprised.

Now figures start pinging round the room. Anyone who wants to debate fishing should be prepared to be swamped by a battery of data from government authorities, the scientific community and the fishing industry. Major economic interests are at stake and many people have an interest in shifting attention away from the real problem, for instance by turning the spotlight from eels as a threatened species to regional policy instead. Swedish policymakers claim that eel fishing supports large sections of the country's small-scale fishing fleet and in doing so is vital for the survival of coastal communities. This is despite the fact that the total eel catch is worth less than four million euros per year in Sweden[1], and declining. And four million is a gross figure; profit after costs is far lower. If you are to believe this reasoning, the survival of coastal communities in Sweden hinges on the price of half a dozen condominium apartments in Stockholm.

And for that amount the state seems prepared to sacrifice an entire species.

Glass eel fishermen perhaps earn a bit more. When the Director-General of the Swedish Board of Fisheries asks what will happen to eel prices in the future when eel catches drop, the man from Glass Eels reluctantly predicts a rise from 870 euros per kilogram in 2005 to maybe 1,300–1,400 euros in the next few years.

It becomes obvious that despite the enormously critical state of the eel species, there is still plenty of money to be made. Some 1,000 euros a kilogram for babies of an acutely endangered species cannot be considered bad business economically speaking, and it is also completely legal. And yet, if we spend a few more years dithering over philosophical issues and turning a blind eye to biological facts, an entire species will be gone. Some researchers fear this could happen at any moment. The Sargasso Sea is large – the surface of the EU 27 countries together, and many thousand metres deep – and scientists believe that the eel population needs to have a certain critical mass to ensure that sufficient numbers of adult eels can actually find each other in the huge water space in order to mate. That critical mass may already be too small.

I now feel my time has come. Two years have passed since I first set eyes on that press release from the Board of Fisheries. It is panel discussion time, the floor is open for the audience, and soon the whole seminar will be over. Willem Dekker from the Netherlands, Cedric

---

1   Throughout the book the simplified exchange rate of 1 euro = 10 Swedish Kronor (SEK) is used. Actually the exchange rate has varied over the years, sometimes 1 euro being 8.8 SEK, sometimes 11 SEK.

Briand from France, Peter Wood from the UK and Kenneth Patterson from the European Union are all sitting on the podium in front of me. The air in the hall is getting stuffy and it smells of coffee. The Swedish Fishermen's Association has already hijacked the event twice by taking the floor unannounced and handing over a raft of data to Patterson. The debate is heavy on detail but low on enthusiasm. And now it's time for me to do it. Now I have to ask the question that no one else has asked all day – the question I have never heard asked in public before; the question I have been turning over in my head for the past two years.

I put my hand up. Patterson nods and I find a crackly microphone in my hand.

"If I've understood things correctly," I say, "eels are critically endangered. So, I can't understand why we're discussing anything other than a total ban on eel fishing here? Doesn't the eel qualify as a species which should be protected?"

Behind me a Green Party adviser titters loudly. The panellists on the other hand all sit in silence, looking strained, as if someone has clamped them with invisible gags. Patterson speaks up, avoiding eye contact with me as he launches into a monologue. It would be very unbalanced, he says, in fact it would be distorting reality to lay the entire blame for the plight of the eel at the door of the fishing industry. Other contributory factors, like pollution, global warming, toxins and disease cannot be excluded and measures need to be taken across all these areas. Just pointing to fishermen would be *completely* unbalanced, he repeats.

"Completely unbalanced," I scribble automatically in my pad as the man from Glass Eels Ltd. suddenly feels the urge to add to Patterson's answer. Wearing his stylish navy blue suit, he rises from his chair, clears his throat and waves his papers in the air.

"Remember," he says, pointing at me, adopting that strange, nervous smile again: "Remember what I said, Isabella. This is your fault too. Remember, it's everyone's fault."

THE ALARM BELLS TOLL

Summer 2001

In 1992 the world's largest cod fishery, off Canada's Newfoundland coast, collapsed. After years of intensive fishing, warnings from scientists, depleted stocks, more efficient catch methods, more public subsidies to fleets, even more depleted stocks and lastly, an unseemly race among cash-strapped fishermen for the last remaining cod, the day came – the fish were finally gone. The mythical fishing grounds, once described by Canada's discoverer, John Cabot, as teeming with metre-long cod so common that they could be hauled up in baskets, had been vacuumed to emptiness after only a few short decades of intensive bottom trawling.

What had we, on the other side of the Atlantic, learnt from this disaster?

Strangely enough, nothing.

Swedish Green Party spokesperson, Maria Wetterstrand, recalls a phone call she made in the summer of 2001 to Lennart Nyman, then head of nature conservancy at WWF, asking if he had seen the European Union's Green Paper on the Common Fisheries Policy (CFP), published in March that year. The tall, slender, silver bearded Nyman, sitting in his office at the Ulriksdal Castle in a Stockholm suburb, confessed immediately that he had only leafed through the document, but had not had time to really read it.

Back then, few people in Sweden had heard of fishing as such as a potential threat to fish stocks. But Wetterstrand had read the Green Paper and had grown more and more upset with each page. Written in dry bureaucratic language, the document had clearly been vetted by numerous officials, every word weighed to avoid unnecessary treading on toes. But the text was still explosive. After 20 years of the CFP, it said, the EU had failed miserably to establish sustainable fisheries and many fish populations were well below biologically sustainable levels. Most worrying was the plight of demersal (bottom-dwelling) species like cod, hake and whiting. The problem, the document continued, was not pollution but fishing. "The available capacity of the Community's fishing fleet is far greater than what is required for fisheries to be sustainable."

The Green Paper also accused politicians of having turned a deaf ear to scientific advice for decades. "The Council has systematically agreed on total allowable catches (TACs) higher than the quotas recommended by the European Commission based on scientific advice."

"I felt there was something there, a political area no one had highlighted," Wetterstrand recalls. "I saw [Swedish film maker] Peter Löfgren's documentary *The Last Cod* on TV later that winter and that reinforced my impression. But I needed to find out more and so I called up a few people like Nyman."

A fisheries scientist, Nyman had spent more than twenty years at the Swedish Board of Fisheries, including a stint as head of its freshwater laboratory, but he had also worked for the Canadian Department of Fisheries and Oceans in the early 1970s. He agreed with Wetterstrand that he was ideally placed to give an assessment of the EU Green Paper on the CFP and what it might mean for Swedish fisheries.

"Fortunately I had some time to do it," Nyman recalls. "And I basically thought it was a surprisingly good and perceptive document. It said all there needed to be said really: that too much public money had been spent on subsidising too many fishing boats chasing not enough fish. It was as simple as that."

Having spent almost his entire working life specialising in aquatic environments, Nyman agrees it was surprising that not even he had the full picture of the state of Europe's seas as shown in the Green Paper. "And I'd already seen it happen in Canada."

Nyman discussed with Wetterstrand the collapse in Canada's cod stocks – once the richest cod fishery in the world, with an annual catch of up to 800,000 tonnes in the late 1960s declining to 250,000 tonnes in the 1980s. By 1992 the cod were gone and fishing was banned. Stocks have yet to recover and the Grand Banks remain closed to cod fishermen.

"We're talking about 40,000 jobs – a cost of billions of dollars to the Canadian government," Nyman observes.

The same thing happened in US waters some years later, at Georges Bank off Cape Cod. A moratorium on fishing saw the cod stocks stage a recovery, but the numbers have never recovered fully because over-intensive fishing irrevocably altered the ocean ecosystem.

"That's what can happen. The species lower in the food chain take over. In the Baltic it's the sprat; off the American east coast it's shellfish and shrimp."

Back in 2000, fishery administrators had only a vague grasp of the so-called ecosystem approach, and quotas were set for each species as if they lived in isolation from the other species in the food web. Sture Hansson, an ecologist at Stockholm University, was a rare voice of reason at that time, insisting on the frightening domino effects there might be if the Baltic cod stocks were to collapse. His theory was that overfishing of cod in the Baltic (where total biomass of cod fell from a peak of 700,000 tonnes in the 1980s to less than a 100,000 tonnes in the early 2000s) was to blame for the immense algal blooms and the reduced water visibility recorded in recent years. The decline of the cod in the sensitive Baltic ecosystem triggered explosive growth in sprat stocks, which in turn led to a massive depletion of the zooplankton on which the sprat feeds, and thereby fuelled uncontrolled growth in the phytoplankton on which zooplankton feed.

A few years later, in 2004, scientists pinpointed the depletion of zooplankton as the reason for reproductive problems being seen also in Baltic pike, perch and roach. The fish fry were starving to death, apparently due to a lack of zooplankton.

While working as the Green Party's adviser to the Swedish Parliamentary Agriculture Committee, Wetterstrand had seen at first hand the partisan treatment of fishing policy issues. Industry representatives were allowed to present their views in Committee, even handing over boxes of langoustines to show their appreciation to the committee members. Scientists, however, were conspicuous by their absence.

Wetterstrand then unearthed a forgotten 1997 report from the Ministry of Finance with the title *Fish and Fraud – on Objectives, Powers and Influence in Fisheries Policy.* The report made a few ripples at the time, but nothing concrete came of it and it sank into obscurity. Four years on, when Wetterstrand read it, she realised little had changed: government subsidies to fishermen had trebled since Sweden joined the EU, while the number of people employed in the industry had fallen dramatically. But, perhaps what upset Wetterstrand the most was the revelation that eighty per cent of all fish caught in Sweden was being used not for food, but for animal feed.

She instantly realised the issue was tailor-made for the Green Party, but first Wetterstrand (now a Green Party MP) needed to find out more about recent developments. During the autumn of 2001 she called together a group of experts: Lennart Nyman of WWF, Sture Hansson from the Department of Systems Ecology at the University of Stockholm, Stewart Thompson of Greenpeace and Klas Hjelm of the Swedish Society for Nature Conservation.

Initially Wetterstrand was most concerned about the acutely endangered status of the Baltic porpoise – a Nordic cousin to the dolphin – and the 21-kilometre drift nets in which porpoises and seabirds often drowned. (These nets were eventually banned in the Baltic Sea in 2008.)

"But as the picture got clearer we knew we couldn't limit the issue to protecting marine mammals," she says. "To get to the root of the problem we had to address fishing."

The meeting participants turned up armed with chilling data about the state not just of Sweden's seas, but of the world's oceans, where commercial fish stocks in many areas had plummeted by up to ninety per cent since industrial fishing began in earnest in the 1950s.

"The thing was that none of us had a complete picture of the situation in Sweden and Swedish waters," Wetterstrand recollects. "As we were talking, we all started to realise just how bad things were. The Swedish Anglers' Association said the same thing, and they're the only organisation apart from the fishing industry that monitors the situation so closely."

That fishing – with its immense impact on the marine ecosystems through the removal of hundreds of thousands of tonnes of living, mating and feeding biomass each year – was far more damaging to Sweden's marine ecology than pollution or eutrophication was an entirely new notion. So was the realisation that Sweden, for all its eco-friendliness and elaborate standards, was no better than other countries when it came to fishing. On the contrary, Sweden was encouraging targeted fishmeal fishing, in practice leading to the marginalisation of small-scale fishermen fishing for human consumption in favour of huge fishmeal trawlers that can decimate local fish stocks with a single haul. Sweden also ensured an exemption from a United Nations and EU ban on drift nets more than 2.5 kilometres long, which meant allowing drift nets of up to 21 kilometres in length – walls of death for seabirds, porpoises and protected wild salmon – being freely used in the Baltic (up until 2008, when they were finally banned).

Sweden also had another exemption, this time from the EU ban on the sale of fatty fish containing such high levels of dioxin that experts classed it as carcinogenic. But the exemption only applied to people, meaning that fish containing high dioxin levels could be consumed by humans, but not used for animal feed!

On top of this, Sweden consistently disregarded scientific advice and supported the fishing of immature fish and fishing in spawning grounds. And, unlike most other countries, Sweden did not have a single marine reserve, not a single cubic metre of sea, where marine creatures could exist without the threat of fishing.

Lennart Nyman and the other experts at the meeting were unanimous: the marine ecosystems in Sweden were seriously out of balance – not primarily because of industrial emissions or agricultural run-off, but due to another industry that had never been called to account, a fisheries industry regulated not by the Ministry of the Environment or the Swedish Environmental Protection Agency, but by the Ministry of Agriculture and Board of Fisheries – as if it were a production industry as any other, with no ecological boundaries.

The Board of Fisheries has its own research and development department, served by three fisheries laboratories in different locations in Sweden, and it knows more about fisheries and marine life than any other institution in Sweden. Yet it failed to impress a vital truth on policymakers: that pumping in government grants and subsidies to the fishing industry does not lead to higher fish catches in the long run – but the reverse. So instead, government policy had continued to fund increasing fishery "efficiency", an increase that for decades had masked the growing strain being put on fish stocks. Newer, faster fishing boats with better sonar and search equipment and heavily subsidised fuel made it possible for fishermen to spend more hours at sea pursuing a steadily shrinking number of fish, but yet landing almost the same amount of catches. State funding for refrigeration tanks made it possible for crews to stay out at sea longer. Generous benefit rules allowed self-employed fishermen to claim unemployment benefit during off-seasons, bad weather and fish shortages, instead of looking for new employment when stocks went down. Fixed intervention prices set by the EU resulted in fish being caught even when there was no market for it, even if concerned customers boycotted it at the fishmongers or supermarket counter, or if fish auction prices were so low it should not normally have been profitable to even catch the fish – then the EU bought it for a "minimum intervention price", and then the catch was sent to landfills. These and various other measures had aided and abetted fishermen to continue the job they least of all should have been helped to do: to go on fishing until stocks collapsed.

The wording of the EU Green Paper on the CFP of 2001 was surprisingly blunt. Fishing, it said, was by far the main cause of death of all food fish species in European waters. Stocks of cod and whiting in the Skagerrak and Kattegat were "seriously overfished", Irish Sea cod stocks were "in crisis" and cod and whiting off the Scottish west coast were "in a critical condition". North Sea herring stocks in the mid-1990s were near extinction and Baltic Sea cod in serious trouble.

At the meeting table in the Swedish parliament in Old town Stockholm where Maria Wetterstrand and her invited guests had gathered, troubled silence fell. Everyone was struck by the evidence, shocked by what they all realised. That the events in Canada ten years earlier were silently being repeated here – in an ecologically, culturally and economically similar nation on the other side of the Atlantic, right under the government's nose. Sweden, one of the world's richest and most eco-friendly countries, had failed to heed the alarm bells.

Or perhaps no one had rung them?

September 2002

It seemed completely unlikely. As a former environmental reporter, even I had spent a decade in blissful ignorance that the problems in the Baltic Sea environment were due to anything other than pollution and eutrophication – the main focus of debate in the early 1990s. Since then, I had focused on other issues and heard or read nothing in the news to tell me otherwise: until that day in September 2002 when the Green Party demanded an immediate ban on cod fishing in the Baltic as its price for supporting the Social Democrats forming a new minority government.

I was not the only person who was surprised that day. Cod stocks under threat? What were they threatened by? I assumed it was due to agricultural run-off and square miles of lifeless seabed, but all of a sudden people were talking about excessive fishing.

My misgivings only increased when the Social Democrats all of a sudden accepted the Green Party's demand with barely a murmur. And it came as a genuine shock when the Board of Fisheries soon after, on the request of the government, presented a report on the consequences of a one year moratorium. The Board backed up what the Greens were saying: that cod stocks were well below safe biological limits. And not just in the Baltic and the east coast but also in the North Sea and along the Swedish west coast, where the situation was even more serious. The Board agreed with the International Council for the Exploration of the Sea (ICES) and the Swedish Environmental Protection Agency, which both called for a moratorium. A one year fishing ban would certainly help cod stocks to recover, it acknowledged. A ban was deemed to be particularly beneficial for stocks in the Kattegat (a + 17 per cent increase predicted) and the Eastern Baltic (+ 8 per cent) and also in the Western Baltic (+ 6 per cent). The only exception was the Skagerrak, where only a 1 per cent increase was foreseen, because the local cod stocks had already been wiped out.

The laconic nature of this admission confounded me. The Swedish Board of Fisheries was the Swedish state authority on fisheries, employing

300 experts and civil servants – some of them sitting on ICES expert panels – and yet it had remained silent about this impending environmental disaster, completely unknown to the Swedish public, for years – behaving as if it had no will or initiative powers of its own, presenting the extremely bad news of the cod only *in response* to a direct question from politicians. The silence was unnerving. Would the Board advise a ban on salmon fishing if someone asked about that, too? What about halibut and angler fish? For all the public knew, even the humble roach might be threatened. Anything was possible.

Still, it was incomprehensible. Major cod stocks below safe biological limits? Local cod stocks already collapsed in the Skagerrak? Such a serious situation does not just arise out of the blue. Why hadn't anyone rung the alarm bells before it was too late? Where were the warnings? Where was the debate? What had the Swedish Environmental Protection Agency (EPA) and the Ministry of the Environment been doing all this time, while this was happening?

I began searching in news archives and combed through government agency websites for information, but found little to enlighten me. The major environmental organisations had plenty of data, though. I read it, and what I read took my breath away. Was this really happening in eco-conscious Sweden in the 2000s, 40 years after Rachel Carson's book *Silent Spring*, 30 years after the UN Stockholm conference on environment, ten years after the UN commitments taken in Rio? It seemed as if I had fallen into an environmental Bizarro World, a parallel universe where everything on the surface was the same; the same authorities, institutions, laws, the same language of environmental sustainability – but where the outcomes of everything decided and thoroughly managed in a democratic fashion the whole time resulted in *destroying* the marine environment, instead of saving it.

So while Swedes spent the 1990s ferrying their washed marmalade jars religiously to the bottle bank, our staple food fish, the cod, had ended up on the International Union for Conservation of Nature's list of threatened species (IUCN) – a list that also included halibut, haddock, spiny dogfish, porpoise and basking shark. The Swedish EPA's Red List added wild salmon, thornback ray, sturgeon, porbeagle and small-spotted cat shark among others. And then there were the species not on the list yet, but known in 2002 to be in serious decline: eel, hake, pollack, whiting, ling, turbot, angler fish, wolffish and, in the Baltic Sea, pike, perch and yes – even roach.

Some of these species are already gone or nearly so. Only a handful of porpoises survives in Swedish waters; the magnificent bluefin tuna, caught in the hundreds of tonnes by Swedish sport fishermen in the early 1960s, hasn't been seen for forty years.

The more I read, the more the questions started piling up: questions about the widespread dumping of fish that are the "wrong" species, undersize or caught in excess of quotas. Swedish fishermen discard between 5 to 20 per cent of all cod, half of all Norway lobsters and 85 per cent of whiting – a crazy waste of resources that the EU, unlike Norway and Iceland for example, officially condones, well – even requires! I read about the industry that makes fishmeal, a product that must be cleansed of dioxins before it can be fed to animals. I read about the endless subsidies to fishermen; construction grants, compensation when boats stand idle, handouts for ship scrapping, tax relief on fuel that makes it economically viable for skippers to burn eight litres of diesel to produce one kilogram of edible Norway lobster. And the biggest question of them all: How, I asked myself, does it all add up economically? Surely a sector so cosseted by taxpayers must be a major employer and income earner? After all, it was being allowed to do something that society had banned all other potentially environmentally damaging industries from doing back in the 1970s – killing fish (without making food out of it), damaging ecosystems and threatening biodiversity.

The figures I found left me none the wiser. Officially, Sweden had around 2,000 professional fishermen, but the Board of Fisheries' data suggested the real figure was a lot lower. Swedish fishermen with a year-round income generally belong to a vocational unemployment insurance scheme, but this had only 1,400 members in 2004. Nor can it be said that fishermen are high earners. The most recent income survey, carried out by the Board of Fisheries, is from 2000 and shows an average annual income of approximately 5,800 euros. And the annual first hand fish catch value of the whole Swedish fishing industry is only around one hundred million euros (1 billion SEK) a year and "value added" a modest 58 million euros (SEK 585 million) according to official figures. Fishing accounts for just two thousandths of Sweden's gross domestic product of approximately 250 billion euros (SEK 2,500 billion in 2006).

By contrast, the horse industry alone earns annual revenue of 1.8 billion euros, according to Ministry of Agriculture figures. Spending on cats, dogs and other domestic animals generated revenue of 65 million euros and moose hunting also earns more than fishing – around 100 million euros per year of value added, according to the Swedish Hunters' Association.

There were far too many question marks here for my liking. How was it possible that policymakers seemed to have suspended rational thought altogether by allowing such a tiny economic sector to more or less on its own whim govern the biggest remaining areas of wild nature in Sweden – that is, the sea? Was I missing something? Maybe scientists had exaggerated the decline of fish stocks? Or maybe there was some obscure but good reason for replacing wild salmon with farmed fish and allowing unique species like basking shark and sturgeon to disappear while emptying the sea of herring? Maybe behind all this there were some sensible, wise political considerations that a simple layperson like me just didn't grasp?

Like any visitor to Bizarro World, I ended up wondering if I was going mad. Maybe I'd missed something glaringly obvious – that the sea is not our common property after all, and that ordinary people should have no say in the matter? Maybe it was right and proper that the fishing industry, organised by the Sveriges Fiskares Riksförbund SFR (incidentally an organisation appointed "Lobbyists of the year 2002" by the Swedish PR-weekly, Resumé), should have custody of our maritime environment, and that it should be supported in exploiting the fish by publicly financed civil servants, researchers and politicians? Maybe 21-kilometre drift nets that kill endangered species and traces of dioxin in human breast milk were nothing to get upset about?

Or were they?

*

On October 11, 2002, P.G. Öjeheim, undersecretary of state at the Ministry of Agriculture, Food and Fisheries, cautioned that a ban on cod fishing would cost the Swedish state 100 million euros. A week later, the Social Democratic Party fisheries spokesman, Jan-Olof Larsson, put the cost at 200 million euros. Another SDP parliamentarian, Kaj Larsson, (a person who twelve months later would take up a job with the Swedish fishermen organisation, SFR) predicted in a newspaper article that a moratorium would completely destroy the Swedish cod industry. Then Hugo Andersson, a former fisheries adviser to the Centre Party who was now employed as a vice president of the SFR, led a fleet of 27 trawlers and sailed to Stockholm to hand in a petition, signed by various political heavyweights including a string of county governors, against a potential ban on cod fishing. The Board of Fisheries published a report estimating the negative impact on the industry at 60 million euros. It then put out a second report forecasting the total effect on all concerned at 70 million euros, assuming no incidental catches of cod in other fisheries were to be allowed at all. Amidst this flood of numbers, the Swedish Environmental Protection Agency raised a point that everyone seemed to have missed: that it was vital to compare estimated losses against the cost to society if cod stocks collapsed entirely.

While all of this was being discussed in Sweden, the bureaucratic mills of the European Union were grinding the question of a Swedish ban on cod fishing in its own mysterious ways. In January the European Commission suddenly came to a conclusion. Sweden could not impose a unilateral cod moratorium – not even for fish within twelve nautical miles of the coast and thus within Swedish territorial waters – since that would be discriminatory against Swedish fishermen. If Danish fishermen were allowed to chase the last cod, so should the Swedes – since all cod stocks were managed by the Common Fisheries Policy.

Yes, that's right: the very same Common Fisheries Policy that the Commission itself had disqualified in its own Green paper just a few months earlier.

Reactions from the Social Democratic Party and the Board of Fisheries seemed borne mostly of relief, while the Green Party, now cooperating with the SDP in parliament, responded with muted grumbles. Ordinary Swedes, still reeling from the news that cod stocks were endangered, were left confused.

I felt strongly that enough was enough. I rang Henrik Svedäng, a scientist and cod expert, and asked him directly: "Listen, this thing about the cod...? Well, *is* it endangered or not?"

The line went silent.

I explained myself. I found it astonishing, I said, completely unlikely, that we would find ourselves in such a situation with so many government agencies involved, with such close monitoring of the fishing industry by researchers. And if things really were that bad, why would so many parliamentarians, fishermen and even EU officials oppose a ban? "Are things really as bad as it is being depicted in the media? And if it really is, why isn't the cod a protected species?"

Svedäng gave a mirthless laugh. "Well, I'm just a researcher... but if I had my way we'd have banned cod fishing a long time ago."

It was my turn to fall silent, and I felt even more confused. "*Just*" a researcher? He was one of the Board of Fisheries' own scientists, one of those who delivers the data that should form a basis for decision making. If he couldn't influence, who could?

"You'll have to ask the politicians," he replied with the same dry laugh. "All we can do is to provide facts about cod stocks. What we know is that coastal cod stocks are virtually gone and that stocks in the Baltic Sea east of Bornholm are not within biologically safe limits. If breeding success is poor for a few years, which could easily happen because spawning depends on a strong inflow of salty water to the Baltic, then stocks could collapse, just like they did in Canada."

The same resigned laugh greeted all my questions – the same dark chuckle I had once heard in East Germany in the 1980s but hoped I would never encounter outside the Iron Curtain.

"Why has no one done anything?"

"Quite. You tell me."

"How long have we known about this?"

"A long time."

"Who's responsible?"

Svedäng went quiet. The laugh, which I would learn to expect from virtually all the scientists I would speak to over the next year or so, did not come this time.

"The politicians," he replied before correcting himself. "We're all responsible. All of us who know."

HARD FACTS

- From 1996-2010, total catches by the EU-27 countries fell by almost 40 per cent, from 8 to 5 million tonnes. Poland, Denmark and Greece have reduced their catches the most in the last decade.

- Swedish cod catches fell from a record 59,500 tonnes in 1984 to the lowest figure since World War II in 2006, when Swedish fishermen failed to meet their quota of 14,000 tonnes, declaring catches of only 11,000 tonnes.

- 75 per cent of commercial fish stocks in European waters are fully fished or overfished (2011). Catches are up to five times higher than sustainable limits. The official scientific body ICES (The international Council of the Exploration of the Seas) advocates a complete ban on the fishing of a number of species and stocks.

- According to a EU Commission Impact Assessment made in 2011, only 8 out of 136 stocks will be at sustainable levels in 2022 if no radical fisheries policy reform takes place.

- Between 2000 and 2006, the EU paid the fishing industry over 906 million euros to scrap old vessels. In the same period the EU total fishing capacity actually increased by on average 3 per cent a year due to building of new vessels, modernisations and technological improvements.

- According to the FAO, each year 7.3 million tonnes of fish is dumped back into the sea dead, because the fish are too small, economically unviable or caught under a quota that has already been filled. Distributed to for instance every Swedish or Somali citizen (9 million people) that would mean 750 kilos per person and year. Or counted as mean western consumption of fish, it would be enough to cover Swedish or Somali needs for thirty-five years.

- In the North-east Atlantic alone 1.3 million tonnes of fish is discarded each year, according to the EU Commission.

- In biomass terms predatory fish such as cod, tuna, salmon or halibut have declined by two thirds in the world's oceans the last hundred years, with the largest decline occurring since industrial fishing started around 50 years ago, according to an expert panel led by professor Villy Christensen, University of British Colombia, Canada (February 2011).

- In 1945, Sweden had 20,000 fishermen living off a total annual catch of 150,000 tonnes of fish; all for human consumption. In 2008, there were less than 2,000 fishermen living off an annual catch of 230,000 tonnes, mostly fish for animal feed.

- On June 10, 2003, professors Olof Lindén, Hans Ackerfors, Rutger Rosenberg, Staffan Ulfstrand, Lars Ove Eriksson, Lennart Persson and Sture Hansson (of Stockholm, Gothenburg, Uppsala, Umeå and Kalmar and the World Maritime Universities) in a debate article in Dagens Nyheter called for those responsible for the fisheries policy and its consequent negligence towards scientific advice, to be hold accountable and prosecuted in court.

- A study of more than 100 fishing regions published in Science magazine in 2006, concluded that if the present trend in the decline of the world's major commercial fish stocks continues at the same pace as it has the last fifty years, all exploited stocks will have collapsed by 2048.

THE SILENCE OF THE ALARMS

The sound that wakes me up puzzles me. I'm lying in almost complete darkness and need several seconds to begin to understand where I am. I must have imagined being woken up by the sound of mewling seagulls and a pounding old-fashioned fishing-vessel motor; but all I can hear now is the steady drone of an engine – it sounds like a bus.

The smell of rubber boots and dusty wall-to-wall carpets fills my nostrils and suddenly I remember: I'm on board the *Ancylus*, a research ship operated by the Swedish Board of Fisheries. It must be around six o'clock in the morning and we should be leaving Glommen Harbour by now, heading out to the open sea. It's early March and chilly inside my tiny, windowless cabin, despite the small heater on the wall. As I try to straighten out the synthetic blanket crumpled up inside my duvet cover, I hear scrambling noises in the narrow corridor outside. The smell of coffee wafts into the cabin, and so does the sound of muffled voices. Everyone else seems to be up already; sound travels easily in this part of the ship with its shiny hollow walls of brown Masonite board. At twenty-four metres long and six metres wide, the *Ancylus* is a spacious vessel. Built in the early 1970s (and with its décor apparently intact) it takes its name from Ancylus Lake, a giant inland waterway which preceded the Baltic Sea 10,000 years ago.

On board are four laboratories, a workshop and day room, galley, mess, shower, toilet and six cabins. The 1,000-horsepower engines guzzle thirty-five litres of fuel per hour – a fact that is noticeable in the smell of fuel oil hanging in the air.

I have asked to join the *Ancylus* to observe test trawls being conducted to identify cod spawning grounds, a new scientific project in a bid to save the fish from extinction. The previous night the research director Anders Svenson, a thin man with a dark moustache, greeted me in his worn-out long johns. Looking very sleepy he assured me I could safely sleep through the next morning, as nothing interesting would happen until we reached our first trawling location, a couple of hours out to sea at Lilla Middelgrund.

The reason I wanted to join this expedition was because I wanted to see with my own eyes the things I had only read about so far. For a layperson, the whole thing seemed almost incomprehensible. Cod stocks on their way to collapse? How could stocks have been allowed to decline to this extreme point despite all the monitoring and supervision by Swedish authorities?

Or perhaps – had it happened without supervision?

It was on board this very ship that researchers, led by Anders Svenson's colleague Henrik Svedäng, had discovered for the first time in 2000 some of the inexplicable truths of what was going on beneath the surface. The *Ancylus* was test trawling at Brofjorden, a small fjord off the town of Lysekil on the Swedish west coast. Researchers had conducted test trawls here from 1923 until 1980, when for unknown reasons the Board of Fisheries suspended them. Svedäng, newly employed at the Board's marine laboratory in Lysekil, had heard anglers and others claim that cod and other demersal species had totally disappeared in the coastal waters of the west coast ten years ago, so he had suggested it was time to investigate again. Not all his Board colleagues supported the idea. Some argued that coastal fisheries were the responsibility of the Coastal Laboratory, not the Marine Laboratory which mainly should be dealing with the open sea. But at the Coastal Laboratory, which was situated by the Baltic coast, colleagues argued they had other things to do, so Svedäng persisted – and finally got the go-ahead from his boss in Gothenburg.

Finally at midday on February 14, 2000, the *Ancylus* reached Brofjorden and the crew set out the trawl, a large conical net with a 7-centimetre mesh and a 35-metre wide opening, from the stern. That done, the monotonously droning motors of *Ancylus* started pushing the ship ahead at the modest speed of 2.5 knots, covering a 3.5 kilometre stretch of seabed. After 45 minutes it was time to haul. The metal chains rattled as they came out of the water; sea gulls appeared in the air, eagerly waiting for their fast food. But the huge green net looked strangely limp as it emerged. And the crew could hardly believe their eyes when the contents were emptied: four cod. Weight: between 30 and 140 grams. In 1974, a similar haul would have boiled and bulged with more than 400 kilograms of cod. In 1980 researchers on average had caught more than 200 kilos of cod an hour in the very same place!

Svedäng and his team repeated the test – three times over, moving to different areas in the fjord. When they did the maths in the evening, they were all left with a surreal feeling. The result: an average catch of 0.4 kilograms of cod per hour.

The entire week the crew continued trawling in the fjord, in the neighbouring fjords and even the area some nautical miles outside of the fjords, but the results were just as disastrous.

Svedäng called off the expedition so the trawl could be checked by a net maker in the port town of Smögen – maybe there was something wrong with the gear? Everything had to be checked and double-checked; every possible scientific variable that could err had to be excluded. But the answer was no. The trawl was in prime condition. Only the cod was not.

Svedäng and his colleagues then expanded the research area to the entire west coast of Sweden during 2000 and 2001. The expedition findings, published in "Cod Project Steps I-III", make extremely depressing reading. Cod had more or less vanished from the Koster islands in the north to the Kullaberg peninsula in the south. Worse than that, other key commercial fish species like pollock, whiting, haddock, saithe, hake and plaice had disappeared too. The odd fish that were still to be found by the coast were immature and not part of the original coastal stocks that had spawned in the area, but thought to be fish from the North Sea looking for food near the coast. The genetically unique local cod stocks that for millennia had inhabited Sweden's coastal waters were now extinct.

Svedäng and his team published another very alarming piece of information in their report: that the average commercial trawl catches *per hour* had dropped by more than 90 per cent since 1982. In other words, fishermen had offset the decline in their standard catch per hour by fishing for a longer time and more intensively; a modern vessel can even pull two trawl nets at a time, or work in pairs pulling a single large trawl net between them. If fishermen hadn't compensated the decline in the stocks by increasing fishing effort, researchers would have seen a much bigger decline in catches at a much earlier stage.

This trend had in fact already been discovered in 1988 by Leif Pihl, now a professor of the Department of Marine Biology at Kristineberg Marine Research Laboratory, but then working as an independent researcher at the Marine Research Laboratory. But the Board of Fisheries didn't want to publish Pihl's findings at the time. This was another of the mysteries I wanted to investigate; why had his research caused such a furore that the Board of Fisheries pulled its name from an already published report?

Håkan Carlstrand, of the Swedish Anglers' Association, had shown me the report, and it was almost comical to see that someone had stuck a white label over the Board of Fisheries' logo on the cover. The Board disassociated itself from the already printed study, Carlstrand explained, after fishermen protested it had used data from their log books. And the Board agreed Pihl had abused the fishermen's trust by publishing their records after being granted permission only to look at them. It was wrong, the Board argued, to use the voluntary information fishermen had provided against them, and doing so could jeopardise future cooperation between industry and scientists.

What the point was of future cooperation with industry if nobody could use any results? This was a question that remained to be answered. As too was the question about what the industry would gain in the long run by trying to conceal the truth that fish stocks had declined more than the catch data suggested.

Many, many questions. Still waiting for answers.

As I lie in my bunk on the *Ancylus*, wondering whether or not to take a seasickness pill, I still have only a rather vague idea about this strange symbiosis between fishermen, the Board of Fisheries and political leaders, long known as the "Iron Triangle". Research director Anders Svenson had recommended that I take a seasickness pill, but I am not so sure. The cabin is now softly rolling from one side to the other in long, almost imperceptible waves, and the sensation is actually quite pleasant. The red warning triangle on the pack of pills from the chemist also puts me off. Now that I'm finally here to see and perceive as much as possible with my own senses, I certainly don't want to feel drowsy.

That something very odd was going on concerning the fishing industry's hold over government agencies and policymakers had started to become very clear to me during these first few months of 2004. As I followed fisheries policy at every twist and turn in the media, I noted no fewer than three peculiar U-turns, pirouettes and semi-halts in policy decisions during just this short time.

The first volte-face concerned a Board of Fisheries' decision, announced with much fanfare, to extend bottom trawling and purse seine fishing prohibitions off the west coast by enlarging the protected zone by two nautical miles in Skagerrak – to protect the small vulnerable fraction of cod stocks still left. An extension of the zone was also announced for the Kattegat. "Trawling will in principle be banned inside the exclusion zone," Håkan Westerberg, a senior Board researcher, told the Swedish news agency *TT* on August 13, 2003. "We will prohibit bottom trawling and purse seine fishing and there will be a total ban on targeted fishing of cod, haddock and pollock during the first quarter of 2004."

The extension of the zone was, in fact, merely a return to the pre-1980s status quo, since the no-trawling-zone had actually been reduced by one to two nautical miles back then. Yet it was now a red rag to the industry, with fishermen claiming it would completely destroy their livelihoods and hit small-scale coastal fisheries hard because, firstly, they claimed, only larger boats were capable of fishing that far out to sea (a nautical mile is 1.8 kilometres) and secondly, because there were no coastal cod stocks left to protect in any case. This latter fact was also used as an argument by Jan-Olof Larsson, fisheries spokesman for the Social Democratic Party, in a debate article in the *Göteborgs-Posten* newspaper on September 15, 2003. He wrote that new restrictions were pointless because "they are just trying to save fish that everyone agrees don't exist anymore".

SFR, the Swedish Fishermen's Association, complained that the decision to extend the protection zone had been taken without enough consultation with them, which seemed to be a particularly sensitive issue, but most importantly they argued it was based on inadequate data. SFR pointed out the fact that the by-catch of cod and other species had not been studied thoroughly enough; that catch composition hadn't been analysed in detail; that the scientists hadn't calculated the number of immature herring; that the breakdown between different fish populations should have been studied more closely; that environmental impact assessments should have been carried out along different stretches of coast; the economic impact on the fishing industry should have been reviewed; the seabed should have been studied in more detail; the mandatory introduction of selectivity grids for shrimp trawlers should have been trialled in full scale research; and spawning grounds, breeding season and reproductive biology of the herring and cod should have been analysed more closely. In other words, the SFR wanted several years of further research.

The organisation was not mollified by the Board's concession of a fairly large string of "special areas" where trawling inside the exclusion zone would be allowed after all, or by the offer of some unspecified "adjustments", alluded to by Christer Skoog, an SDP member of parliament and a member of the Board of Fisheries' supervisory board, when interviewed in the *Dagens Nyheter* newspaper on August 8, 2003. In fact Skoog criticised the Board's own management policy, predicting that it "would hit coastal fisheries very hard".

The SFR organisation declared war on the Board of Fisheries. The weapon was to ban Board personnel from their boats indefinitely.

Still, the new no-trawl zone came into effect on January 1, 2004. But mysteriously enough, just a few days later more than 30 trawlers could be found trawling inside the protected zone – thanks to various exemptions from the new rules. For instance some of the largest Gothenburg vessels had gained dispensation to fish for herring using lights (a method whereby the fish are attracted by strong lamps at the surface) and purse seine nets at a depth of 15 metres inside the zone.

Why all these dispensations? The Swedish Anglers' Association wrote an open, very critical letter to the Director General of the Board of Fisheries, noting that the only formal basis for the exemptions was the wording in Sweden's fisheries regulations allowing exemptions "if they are acceptable from a fish conservation perspective and other reasons exist". The Anglers' Association noted that the Board had only recently published research revealing extensive by-catch of whiting, cod, pollock and saithe in coastal fishing. So what had changed since then? What were the new reasons that were "acceptable from a fish conservation perspective"?

Karl-Olov Öster, Director of the Board of Fisheries at the time, gave an astonishingly frank answer to the open letter. In his reply he mentioned no new scientific reasons, relying instead on the "other reasons" mentioned in the regulations. These, he said, primarily meant "supply of raw material to the Swedish processing industry".

In other words, industry's demand for fish, in practice, took precedence over conservation. As long as processing companies needed fish to make fish fingers, preserving immature cod had to play second fiddle – a quintessential Catch-22!

But the exemption applications sent in to the Board contained little to suggest that the applicants understood what justifications were needed to gain a dispensation. Only one out of 36 applications mentioned the need to supply fish to the processing industry. The other fishermen's main argument instead was that they had "always" trawled inside the areas concerned. Sure enough, the few rejected applications came from trawler owners with no track record of fishing in the designated zones. Many of the rejection letters read, "On the grounds that you have not conducted trawling [in this area] in recent years, the Board of Fisheries finds no reason to grant a special licence."

Meanwhile, applications to use nets smaller than the minimum mesh size of 16 millimetres (try measuring 16 millimetres between your thumb and index finger for a moment!) gave no concrete reasons whatsoever – and nor did the Board in its responses. A number of the permit applications seemed not to have made efforts to prove their case at all, but seemed merely to be copies of the same text, all with identical spelling errors like "excemption" and "innside the trawling zone". All were granted exemptions from the rule between January 1, and December 31, 2004.

The second U-turn was far worse from a fish conservation perspective, but did not produce a single "open letter" or make the debate sections of any Swedish media at all. As for me, I was watching the television news on February 24 and scarcely had time to shush the children before the presenter had moved on to the next item. I thought I had misheard the snippet, only to double-check on teletext and see it there in black and white. An extraordinary meeting of Baltic Sea fishery ministers had just *increased* Baltic cod quotas for 2004. At their initial meeting in Vilnius in October 2003, the ministers had flouted scientific advice from ICES calling for a moratorium on cod fishing in the Baltic due to uncertainty over the health of stocks, instead agreeing a modest reduction in the 2004 quota from 75,000 tonnes (in 2003) to 61,600 tonnes. Ann-Christin Nyqvist, the Swedish agriculture minister, professed her disappointment to the Swedish media the following day. Sweden, she said, wanted to follow the scientists' advice, but the level of understanding of the dangerous situation of the cod stocks was low in the other Baltic Sea states, she said. The one bright point, according to her, was that ministers had requested further information from ICES and had scheduled a further meeting to consider reducing quotas if new data emerged.

Joakim Ollén, chairman of the Swedish Anglers' Association, expressed pessimism over the excessively large quota, predicting that it would be difficult for the Baltic Sea countries to accept a reduction at their next meeting the following year "as part of the 2004 quota would already have been filled".

Interestingly enough, the fishing industry took the opposite tack, with SFR deputy chairman Bertil Adolfsson briefing reporters that it would be "tricky" to live with the lower quota, but predicting that the new ICES figures requested by the fisheries ministers would "show that there are more cod in the Baltic than the biologists think".

This was a strange statement, knowing that the biologists he thought had got it wrong were the same ICES biologists that were supposed to come in with the new data. But four months later Adolfsson was indeed proved right: rather than cutting cod quotas, the ministers raised it to 75,000 tonnes – the same figure as 2003.

So – were there any new data to show the state of cod stocks weren't as bad as presumed? Not at all. ICES experts were still advocating a moratorium on cod fishing in the Baltic that year. In their opinion, the maximum "take" from cod stocks east of the island of Bornholm was 13,000 tonnes, the same figure they proposed in October the previous year. ICES also highlighted all the unknown variables making it impossible to measure the exact biomass of Baltic cod. Unreported landings (illegal fishing) and discards (dumping) were such variables – and in fact these were the only figures that differed from the data presented to the first ministers' meeting in Vilnius. This time, without any explanation, the estimate for unreported landings and discards had been reduced by more than 70 per cent, from 52,600 tonnes to 14,600 tonnes.

I call Robin Rosenkrantz, chief negotiator of the Swedish Ministry of Agriculture, Fisheries and Food, to ask how this can have happened. He has a difficult moment on the phone trying to express disappointment at Sweden's failure to persuade neighbouring countries to reduce the quota, while at the same time explaining that the increase is quite reasonable and poses no immediate threat to cod stocks at all. It's simply a question of *time*, he explains. ICES scientists believe a moratorium is the best way to ensure *the quickest possible* recovery in cod stocks (which are now below the 160,000 tonne limit regarded as constituting critical biomass and far below the limit of 240,000 tonnes required under the so-called precautionary principle).

"ICES only takes biological considerations into account," Rosenkrantz says. "Their job is to tell us that the longer we are below the safe limit, the greater the risk we are taking."

But, he stresses, the government takes a different view. Politicians are "managers", and managers must take other factors into account too, such as socio-economic considerations. It is these that have led to governments taking a different approach.

"And with the new ICES estimate of total biomass it means we can stick to our political objective of increasing biomass by 30 per cent per year until we have reached a safe level."

I don't follow him. What new biomass estimate is he talking about? I haven't seen any new estimate on biomass?

Well, Rosenkrantz explains, with the downward revision in the figures of illegal fishing and dumping, we must assume that we now have 38,000 extra tonnes of living and swimming cod in the Baltic – an increase in numbers that will enable the entire stock to multiply enough to reach the precautionary principle target of spawning stock biomass within just a few years. And, he concludes, this is *exactly* the political goal established by the Baltic Sea governments and endorsed by Sweden!

Rosenkrantz seems happy. At the end of the day, everything comes within approved frameworks of targets and political agreements; nothing to worry about, he assures me.

"Only a question of time," he says.

Afterwards I speak to someone close to ICES and ask what really led the council to reduce its estimate of illegal fishing. Was there any new data? The answer was no. The first estimate was based on information received from the coastguards and some fish landing ports, sources that all want to remain anonymous. The figures for discards were more reliable, but none of them were really hard, verifiable data. So the ICES representative was simply questioned and pressed so hard by the Baltic Sea ministers that he finally agreed to cut the estimate by 38,000 tonnes – more than 70 per cent – right at the negotiating table.

Thus, the life conditions for an apparently unknown number of cod, genetically unique on account of their ability to live in brackish water, were transformed at the stroke of a pen. A population of cod that has just one remaining spawning ground, a variant of the species whose future hinges on future autumn storms being strong enough to force oxygenated salty water through the Öresund Strait and into the Baltic so that the cod fry will survive. A fish reliant on reduced predation of its young by shoals of sprats, and on its young finding sufficient quantities of zooplankton on which to feed. A fish dependent on no new cod disease occuring (or any mysterious disease like M74 that plagued wild salmon stocks) as long as the stock is outside safe biological limits.

In other words, a fish that no one knows how much time it has left.

A watery morning sunshine attempts to pierce the thin layer of clouds. The sea is steely grey, monotonous and a world away from any touristic glittering boat trip memory I can produce in my mind. Remorselessly, the *Ancylus* ploughs on, sounding like a bus in the greyness, past the no-trawling zone and into European Union waters shared by Sweden and Denmark.

There are five members of the crew, including the two scientists, and they are all men. I say good morning to them and return back down the steep narrow staircase to the mess, not missing the sea view. The quarters down below are a dingy brown, with yellow sliding cupboard doors and a guard on the cooker to stop pots and pans from sliding off. A breakfast programme is showing on the television mounted on the wall, with people talking and talking, sitting on sofas. The pantry and fridge are well stocked and I make myself some tea and sandwiches for breakfast, while I try to concentrate on what is being debated by the groomed figures on the screen. It all seems strangely distant and irrelevant though, as if only a few nautical miles out to sea on this boat I am already in another world.

I sit down at a shiny imitation-wood table that is bolted to the floor, and stare at the framed pictures of boats hanging on the walls. The sea is still smooth.

Researcher Anders Svenson enters the quarters, now dressed in orange oilskins with a navy blue woollen hat pulled down tightly over his ears. He tells me that I have been lucky with the weather and may not have to worry about being seasick after all. This is the third day of the expedition and the calmest weather so far. Anders explains that the ship has criss-crossed the Kattegat, starting in Höganäs, and will end in Göteborg. So far they have identified a handful of potential cod spawning grounds. Oddly, the crew had found sexually mature fish just 25 centimetres long. This is strange because a cod – which can grow to two metres if it is not prematurely killed – usually reaches reproductive age at around 40 centimetres long. Exact length varies between regions and stocks though, but suffice to say that Baltic cod are sexually mature at around this size (about three years of age), while cod off the Swedish west coast are larger still, perhaps 45 centimetres, and Icelandic and Canadian cod even bigger. These are important facts in a fisheries management context because no one claims to want to fish for cod that haven't yet spawned once. Killing cod "teenagers" that have yet to mate and reproduce is obviously pure extermination policy. This is why there are minimum landing size limits for catchable cod. But these are, strangely enough, some way *below* spawning size – 38 centimetres in the Baltic and 30 – 35 centimetres on the west coast. The Anglers' association has campaigned for decades for their self-imposed limit of 45 – 50 centimetres to be enforced also in the professional fishing industry, but in vain. It must also be added that some researchers emphasise that the most reproductive fish are the old ones who are able to spawn successfully year after year and therefore efforts should also focus on conserving big fish and not only the young-sters: *maximum* landing sizes would be perhaps also be appropriate.

But might it not, I ask Anders, be a good thing for cod to reach sexual maturity at just 25 centimetres? Couldn't that perhaps be the salvation of the fish? He looks sceptical: No, abnormally small reproductive cod are *not* good news – this was exactly what happened in Canada shortly before the cod stocks collapsed. Unless the fish are some kind of stunted variant, he says, the phenomenon suggests the species is under serious stress. Earlier sexual maturity may be nature's own way of trying to preserve the species but is unlikely to succeed without a radical reduction in fishing. As the Canadian experience demonstrates, not even a mora-torium may be enough.

I remember a line from a British book about how young cod follow their parents to the spawning grounds in their first year to learn how to breed, and ask Anders if this is true. He has not heard this himself and wants to know more. Even if he is an expert, there are lots of things he doesn't know. The fact is that cod occur in different popula-tions throughout the North Atlantic and all have their own habits and patterns of behaviour. The only thing we know for sure is that we simply do not know enough about them.

We are approaching the first trawling area and Anders goes back up on deck. I stay below pondering the fact that we seem to know more about life on Mars than life in our oceans. As late as the 1990s the Norwegian oil company Statoil found a previously undiscovered 13-kilometre long coral reef with white and orange-red *Lophelia pertusa* in deep water off northern Norway. Not until comparatively recently have scientists realised that coral reefs occur not just in tropical water but in cold, Nordic environments too. The coelacanth, thought to have been extinct for millions of years, was rediscovered accidently in 1938 by a museum curator among other fish catches landed on a South African quayside. The Mariana Trench, the deepest point of the world's oceans, 11,034 metres below sea level, has been visited by humans only once; by Jacques Piccard and Don Walsh in 1960. The interest in exploring the 1.4 billion cubic metres of water surrounding our planet, accounting for 71 per cent of the Earth's surface (and an even larger percentage of the world's biosphere when you consider that the ocean is three-dimensional) seems almost non-existent. In 2004, when I asked a Swedish marine biologist specialising in eel research why, in a moment in history when we can send space probes to Mars, we cannot tag eels with GPS transmitters to find their mystical spawning grounds in the Sargasso Sea – he countered in a surprised tone of voice. "But sure we can! It's not impossible, only quite expensive." It's a lot easier to raise money for eel fishing than for eel research, he noted.

No doubt he was right. During the 1990s Sweden spent between 200,000 and 800,000 euros per year on eel restocking in order to prop up the eel fishing industry. In some instances, this so-called "fish conservation" money benefited eel fishermen twice, once by giving money to fishermen on the west coast of Sweden for catching living young yellow eels – which are then released on the east coast to be caught by other fishermen a few years later, when they turn into silver eels. Among the new boats built with the help of EU subsidies from 1995 to 2004 – a period that saw a catastrophic decline in eel stocks across Europe – we find four new Swedish eel fishing vessels (three boats and a barge), all built with large subsidies of taxpayers' money.

We've already touched on the third strange pirouette – the use of drift nets. Rarely has such a peculiar aspect of Swedish environmental policy received so little media attention. How many people know that the Baltic fishing fleet has been using a technique rejected by the international community as far back as 1991? How many of us know that drift nets up to 21 kilometres could be used in the Baltic, and that the EU has for years been pressuring Sweden to cease this indiscriminate drift net fishing of salmon because it threatens the tiny remaining population of porpoises that still survives in the southern Baltic?

In 2003 the EU proposed an interim cut in drift net length from 21 kilometres to 2.5 kilometres from July 1, 2004, with an outright ban from January 1, 2007. But Sweden opposed the plan, with Håkan Westerberg, of the Board of Fisheries, claiming that the "50 to 100 Swedish boats that would be affected in some way" needed "more time" to adapt to the change.

Negotiations in March 2004 delivered a deal to reduce fishing gradually by fixed annual percentages over five years. In practice, this allowed Swedish fishermen to continue using 21-kilometre drift nets until January 1, 2008 – 16 years after the United Nations agreed an international moratorium, 6 years after the EU imposed its own ban and 5 years after the European Commission announced proposals to ban them in the Baltic Sea.

Such a generously long transition period is even harder to understand when one considers that scientists and other experts have long insisted that selective salmon fishing without drift nets would work fine if conducted at river mouths frequented by restocked salmon. (In Sweden, all the hydropower companies are obliged to restock salmon in rivers that have been blocked by water turbines.) This would radically reduce the number of seabirds snared in drift nets (one survey found that 34 per cent of all ringed recoveries of guillemots were drift net fatalities), conserve the threatened wild salmon and remove one of the threats hanging over Sweden's only remaining small whale, the common porpoise.

Despite winning such a long transitional period, the fishing industry lamented the new restrictions on salmon fishing with emotive coverage in the *Yrkesfiskaren* trade magazine in early 2004. "Death penalty on the drift net fishing," they wrote; "a black day", the fishery is approaching its "definitive end" and "extermination".

But according to one source at the Board of Fisheries, fishermen were already planning to solve the problem by switching to longlining with baited hooks – an indiscriminate method of fishing that has earned worldwide criticism for killing even more birds than drift nets.

Anders and crew member Peter are setting the trawl net for the first tow. The two heavy metal trawl doors are first lowered into the sea from the stern. They plough along the seabed and are angled outwards in order to expand the opening of the net to a maximum. A number of floats hold the upper edge of the net open, while weights hold down the bottom edge. A footrope consisting of rolling balls ensures the net moves smoothly along the bottom. The trawl "bag" is around 25 metres long and 10-12 metres wide. I have seen underwater film footage showing what happens on the seabed when the trawl net approaches. The trawl boards scare bottom-living flatfish like flounder and dab into taking flight, though they do not get very far before being swept into the net. Cod put up more of a fight. Oddly enough, perhaps because of the chaos created by the trawl boards with the sound and whirling sediment in the water, they don't attempt to swim beyond the trawl boards to escape, but instead swim steadily straight ahead, giving it all, letting themselves be hunted forward metre by metre. Famed for its white, tender flesh, the cod is more of a sprinter than a long-distance runner, as befits a predator. So, after its final explosive spurt it becomes completely exhausted and is sucked into the net. So exhausted are the fish that smaller individuals then often lack the energy to find the escape window that some trawler nets have to allow undersized fish to swim free.

And undersize fish that fail to find the escape windows are all likely to die. Either they sustain fatal injuries from ascending too quickly when the net is hauled in, the fish version of decompression sickness (it is not uncommon to see fish coming out of the nets with their stomachs protruding from their mouths due to an overinflated swim bladder). Or else they die of injuries or lacerations sustained when tumbling around inside the trawl net during the tow. The few fish lucky enough to remain unscathed on deck must then run the gauntlet of an army of seabirds precision-diving to snap them up when they are thrown back into the sea. Mortality of discarded fish (in fisheries bureaucracy language the euphemistic term is "discarding", not "dumping") is between 90 and 99 per cent – and the number of unintended, unwanted, undersized cod that is caught each year is indeed extremely high. Professor Kerstin Johannesson of Tjärnö Marine Biological Laboratory estimates that only 1 million individual cod of the 17 million caught off the Swedish west coast every year have a marketable size; the rest are discarded. Sixteen million individual fish that should have been allowed to grow and mature thus end their days as food for seagulls.

Anders decides to limit the tow to 30 minutes, partly to avoid causing unnecessary suffering to the fish. We are in classic fishing grounds, between Anholt and Läsö, but there are still no other trawlers in sight.

"We're only in early March and 66 per cent of the cod quota has already been filled," Anders explains. I look round. The concrete grey sea is now turning dull blue, the sun is out and the wind is so strong it makes my eyes tear. We have only been two hours out at sea, but already there is no coastline to be seen in any direction. Out here it is easy to understand what astronauts so often talk about: that the Earth, when seen from the heavens, looks all blue – and that the orb that we humans with typical anthropocentrism call "Earth", "Terra", "Erde", really ought to be named Ocean. Water covers almost three quarters of the planet's surface. From most positions on "Earth" it is just like here; you see no earth at all, only mile upon mile of blue sea, home to billions and billions of marine organisms. In fact, it was here in the big blue that life once began. It was here, in this primordial soup, that biochemistry once ignited the first sparkle of life. Here phytoplankton produce more oxygen than the world's entire rainforests. Here the coelacanth developed the lobed fins that enabled its descendants to clamber on to land. And we humans still carry the remembrance of the sea within us, e.g. in the salinity of the liquids in our cells, and the fin-like appendages which, in foetuses, precede the formation of hands.

My thoughts turn to an article in *Nature* from May 2003 in which two well known marine biologists, Boris Worm and (the since deceased) Ransom Myers, compiled all the known data on the major global commercial fish species over the last 50 years. All these stocks have declined by 70 to 90 per cent, according to Worm and Myers, due to an industry that has grown too effective for its own good. In just fifty years – thanks to sonar, satellite navigation, drift nets, longlines, electric winches, modern refrigeration and aerial fish school spotting – humans have succeeded in radically and fundamentally disturbing a marine ecological balance established over many millions of years. We still know very little of the long-term consequences of such a dramatic shift. In some places we have seen that certain fish species have taken over from those that have vanished, entire food chains have been disrupted, algae have invaded previously clear water, and jellyfish and shellfish have increased exponentially, as on the Grand Banks. The common denominator for these phenomena is that we humans have removed the predators at the top of the food chain – cod, halibut, salmon, whales, dolphins, sharks, tuna and sturgeon – and left the field clear for what the biologists call the "opportunistic species". If we translate it into terrestrial terms: it's like culling all our wolves, bears and foxes and leaving deer, hares, rats and other rodents as masters of the environment. To continue the analogy: we are now putting these hares, rodents and deer under too much pressure; we are "fishing down the food web". And we simply do not know yet how far down the chain we can get before the oceans are awash with jellyfish and plankton with no natural enemies, or when zooplankton communities collapse because of predation by opportunistic species: where the breaking points are such that entire food webs disintegrate.

What we do know is that zooplankton stocks are alarmingly low in the Baltic. An unchecked increase in sprat and herring stocks due to the overfishing of the cod is thought to be responsible. Competition for food among the increasing number of sprat and herring has become intensive. In the last decade, average herring weight has fallen by forty per cent and the species also carries dramatically much less fat than in the past. Even more worryingly, researchers suspect that this may also be affecting predators that eat sprat and herring, such as salmon and seabirds. There is no demonstrated link as yet, but it is worth noting that the mysterious salmon disease M74 was due to vitamin B deficiency rather than a virus or pollution, as first believed. The cause of the deficiency remains unclear, and research fell into abeyance when scientists discovered that feeding the vitamin B supplement thiamine to farmed salmon protected them from the disease.

The need for more research on this strange vitamin deficiency became obvious in the summer of 2004 when mass mortality among seabirds in the Baltic – up to 80 per cent in some colonies – was found not to have been caused by some pollutant or mystical virus, as first believed, but rather by an acute vitamin B deficiency. The same summer, the Board of Fisheries' marine laboratory analysed rising mortality among young pike, perch and roach off the Stockholm coast and discovered that the young fish had severe problems with finding food near the coast. The reproductive problems that had been noted over a number of years, which had led to declines in fisheries of those species, and even a fishing ban in a very large area along the southern east coast of Sweden, had been a mystery so far, since scientists had observed the adult fish spawn, and the fry to hatch properly. Now they found the answer to why the fry didn't survive: once they had consumed their yoke sack (like the fish "placenta" that it is born with), they starved to death because they couldn't find enough zooplankton to feed on. The sprat and herring had eaten all the zooplankton.

I thought we were fishing for cod, so I am very surprised to see all the other different species that appear when the net is reeled in after half an hour. Fifty kilograms or so of sand dabs, lemon sole, hake, whiting, capelin and herring. And a gurnard!

"Watch out for that one, it's poisonous," Anders warns. The shiny little frog-like fish with its dragon-like dorsal fin can deliver venom that is somewhere in-between a wasp sting and an adder bite.

But there are some cod too. Quite a lot, I think initially with relief, though I realise I have no idea what a "normal" trawl would bring in. This cod catch comes in at a combined weight of 16 kilograms, distributed among 30 fish.

"Not so bad!" I judge.

Scientist Hans Hallbäck shakes his head sadly. "Here we are in the middle of the spawning season in a well known fishing ground," he says. "We should have caught maybe a hundred kilos of adult cod here, but virtually all we've got are one-year-olds. If the fish stock is healthy, you catch a range of ages. This only confirms just how bad things are."

Anders and Peter weigh and sort the fish by species, recording the data in notebooks. The by-catch is then thrown back into the sea, and the gulls dive like torpedoes into the water after the boat, small flounders flapping in their beaks, silvery mackerel and whiting flashing in the sun. The gulls that manage to snaffle the larger fish fly off to the horizon to escape their comrades' raucous cries and attempted piracy. From what I can see, not a single fish escapes the birds, apart – possibly – from the gurnard. The gulls' perception skills seem almost miraculous.

"Did you notice there wasn't a single gull in sight before we started hauling in the trawl?", Bosse, the friendly first mate, himself an ex-fishermen, asks. "They turned up as soon as we started using the winch. I've always wondered how they do it," he says, looking genuinely puzzled.

First each cod has to be measured. Then deckhand Peter quickly slices them open and removes their reproductive organs. These are placed on a stainless steel scale to be weighed and recorded. The weight reveals whether the fish are sexually mature or not.

This particular tow has yielded only a few spawners and Anders concludes that it is not a spawning ground. Spawning, he says, starts with the male establishing a territory of a few square metres on the seabed where he lies in wait for a female. If a passing female likes the look of him, she will approach and pick him up for a courtship dance. They move together upwards, in spirals, clinging belly to belly. At a certain moment the female releases her roe and the male his sperm. Research shows that older, more experienced individuals have the best reproductive success rate. Old females are especially important because they can carry millions of eggs – at least ten times more than a first-time spawner.

If everything has been successful so far, the fertilised eggs drift in a special water layer between the surface waters and the cold bottom waters, which has exactly the right salinity which the eggs depend on. Unfortunately, here they are at risk from predators. If cod stocks are low it tends to lead to an increase in numbers of, for example, herring or shellfish, which feed on cod eggs and larvae – perhaps one reason why the cod has not recovered in Canada despite the fishing moratorium.

Once the fry find their way to the coast and sheltered bays they have better survival prospects. Here they bide their time and grow, though the environment is rarely safe. Like herring, young cod are readily drawn to night-time lights shone from purse seiners at sea in the middle of the night. Giving protection to these young fish was exactly the reason why the no-trawl and purse seining zone was extended farther offshore.

Anders says he is surprised at all the exemptions given to trawlers by the Board. They fly in the face of the scientists' advice, he says with a resigned shrug. But, he adds with an odd smile, experts at the Board's Marine Laboratory are used to their employer treating their research with a pinch of salt, as if it were a stakeholder's view to be balanced against other interests and not objective science. When the Board recently restructured its organisation this view was emphasised. The responsibility for the Marine Laboratory shifted to a special "resource department" at the Board's Gothenburg office – a department run not by biologists but by administrators whose job involves weighing the institute's research against – well, of course: socio-economic factors.

These socio-economic factors were obviously in the mix when the exemptions were handed out. But now good news is that the herring fishing inside the trawling exclusion zone has resulted in much lower incidental by-catch of cod than experts predicted. The news is in today's Gothenburg newspaper *Göteborgs-Posten*. One fisherman tells the news-paper that he can't remember herring fishing being as "clean" as it is now, and that the herring are well nourished and occur in dense, silvery shoals, just like in the good old days. The scientists had once again been proved wrong, he claims. This irritates Anders.

"The by-catch of cod is lower because there soon won't be any cod left!" he says while shaking his head with the woollen navy blue hat pulled down over his eyebrows. He cleans the fish blood from the gutting table and adds tersely, "And they're fishing the last of them, right in the nursery."

There is actually one bright spot for Swedish cod fisheries, believe it or not. It is the Öresund Strait. Here, in the narrow bottleneck between Helsingborg in Sweden and Helsingör on the Danish side, which is criss-crossed by intensive ferry traffic and spanned by the controversial Öresund Bridge, here in one of the most densely populated regions in Scandinavia, with Barsebäck nuclear plant emitting hot water and where Scandinavia's most productive agricultural areas have their run-off, where the fishing fleets on both sides operate intensively under different rules on minimum fish size and legal fishing periods – here, strange as it may seem, cod and other fish stocks are faring rather well. Hourly cod catches are at least ten times higher than in the Kattegat – and the fish are larger too. Fish of less than 20 centimetres predominate in the Kattegat, while fish in the Öresund Strait are more than double the size, 60 – 80 centimetres in length on average.

The Öresund example is inconvenient for all the people who blame emissions, pollution and marine transportation for the cod's demise.

The strait has all these pressures – considerably more so than the rest of Sweden's coasts. Why the apparent paradox? The answer is that urbanisation actually saved the fish.

Ever since 1932, authorities have banned all types of fishing with towed gear, including trawling and purse seine fishing, since they are deemed potentially hazardous to the intense maritime traffic. And ever since it seems cod, haddock, whiting, lemon sole, plaice and other bottom-dwelling species have found something of a haven in Öresund, despite extensive fishing with fixed nets and rods (and rumoured illegal trawling). This suggests that fishing with fixed nets with a certain mesh size that only target fish in a certain size interval (which allows small fish to pass through the mesh while the largest fish don't get entangled, but bounce off the net), is a much more sustainable method than trawling. As we now know, the largest fish are by far the best breeders. Conserving them might be equivalent to saving the goose that lays the golden egg.

The equation may look pretty simple, but as the abnormally skewed curves showing cod sizes in the Kattegat show, it has not yet sunk in with fishermen or with the "resource department" of the Board of Fisheries.

On the upper deck notice board I spot a "Wanted" poster, one of hundreds sent out to Swedish fishermen asking them to watch out for cod with a small greyish cylinder marked "Board of Fisheries" attached to one of their fins. Scientists are tracking the tagged fish to find out more about their movements. Until recently, some experts believed that cod were so mobile that it was almost pointless to ban fishing in marine protected areas or reserves. Now, however, there are indications that some cod populations are fairly sedentary. Cod in the Öresund Strait are an example, and researchers claim that the disappearance of the cod along the Swedish west coast proves there were once stationary coastal stocks that have now become extinct. The suggestion is that the small individuals currently being caught off the west coast are only visiting immature cod from stocks farther out in the North Sea.

Anyone who finds a tagged fish and sends the tag to the Board of Fisheries with information on the time and place of capture receives a reward of 80 euros. That's the only way, Anders chuckles ironically, because without a reward the Board would not receive a single tag due to the ongoing conflict between the Board and fishermen over the no-trawling zones. But now numerous tags have already been received: west coast cod do not evade capture for long. "We get all the tags back. The record is in two days," Anders says.

Peter comments, talking about the conflict.

"We scientists aren't popular at all right now after the dispute over exclusion zones. If you talk to fishermen you always have to say 'I'm not like all the others. Personally I understand what you're going through.' But you still don't really want to be walking round with a jacket saying Swedish Board of Fisheries on your back."

Now the fishermen's boycott of Board staff has been going on for more than two months, causing the postponement of research into by-catch and discards, as well as delaying trials of more selective fishing methods.

But, I ask, how are the fishermen allowed to do this? Surely it must be possible to change some rules and *require* skippers to allow Board personnel on board? After all, farmers cannot prevent inspectors from entering their farms!

Anders and Peter explain that there is a difference between controls, which are carried out by the coastguards, and the Board's research, which relies on voluntary access. "And anyway," one of them comments, "I promise you this: you *really* don't want to be on board a boat where you're not welcome."

The conversation moves on to what happens if Board officials discover misconduct on board a boat. "Like that skipper that logged a catch of capelin as dogfish because the quota was full. Or that other that had caught undersize nephrops and kept them ..."

Taken aback, I ask the crew members if such things really happen right under their noses. Absolutely, they say. But then they disagree on whether it is their job to report infringements or not.

"But that's not the reason we're there," one says.

Another contradicts, "That is not an excuse. It is as if you're a police officer at a private party where they're serving illicit spirits, in which case you can't just ignore it and say you're not on duty. You'd be convicted in court if you did."

"But if we reported everything we saw we'd never be allowed on board," a third person chips in.

I can hardly believe my ears. "But whose rules are we really playing by here?" I ask.

"Well, the boats belong to the fishermen. It's their private property," replies one of the crew.

I feel the anger rising inside me, but refrain from asking the questions I frequently ask myself more and more. But who owns the fish? Who does the sea really belong to?

I'm on deck in a blustery wind trying to ignore the fact that the fish thrown into different trays and buckets around me are so obviously alive. They squirm on the sorting table, the cod wriggling their small pointed beards and the flounders curling up in spasms. I pick up a cod that has landed on the deck besides a bucket, and gently squeeze it. Its flesh is soft, cold, slimy, just as it should be. Its translucent, pinkish stomach bubbles out through its mouth. Its eyes are surprisingly beautiful – circular, jet black in the middle and surrounded by a shining golden metallic iris. Bending down, I place it in the bucket and stare at the glaucous pattern on its sides, trying to stop myself from wondering if it can see me too. I know fish have a wide field of view and that their eyes have a similar structure to the human eye.

I think of *Finding Nemo*, the Disney film about the small clownfish that got separated from its over-protective father. The film attempts to portray ocean life accurately – clownfish are certainly paternal and down in the deep there are indeed fish with shining lanterns – and the *Finding Nemo* animators worked closely with marine biologists to ensure the anatomy and movements of their fish would be as realistic as possible. But one realistic detail they could not accept: that real fish don't have eyelids. For the Disney film creators, this was taking realism one step too far. Without eyelids, Nemo wasn't cute any longer – he became staring and soulless.

Anders and Peter have resumed their work of slicing open the cod to take their reproductive organs out and I cannot help but think that maybe we would have treated the fish differently if only they had had eyelids. Maybe it is the absence of this anatomical detail (due to fish not needing them to keep their eyes moist), rather than the fact that they are not mammals, that makes us see them as "just fish". Dolphins and whales have eyelids and they have always captivated us humans. I have myself seen dolphins close their eyes with pleasure when being stroked. And I have seen fish in aquaria, coming on a command, knowing exactly when to expect food. One scientist at a cod breeding laboratory described the species' ability to learn as "similar to dogs".

The swell is getting bigger and I try to concentrate on something other than the fish slaughter taking place in front of me. I ask Anders about the ultimate purpose of this fact finding expedition. When they identify cod spawning grounds, is the aim to protect those areas from fishing?

"No, no, there are no plans on those lines," he replies.

The answer surprises me. But wouldn't that be a good idea? I wonder.

"Absolutely, but it's totally unrealistic. We're not even in Swedish waters. Once we're 12 nautical miles offshore then the EU takes the decisions. And we don't agree with the Danes, anyway. It just won't happen."

As I allow the umpteenth unsettling comment today to sink in, I ask what I imagine to be quite general and harmless questions.

"On what basis have you decided to conduct test trawls in these areas? What makes you guess these are spawning grounds?"

I expect some answer full of scientific data, sea temperatures, monitored seabeds, tagged fish and egg surveys, but to my surprise I hear Anders say:

"Well, we simply visit the places recorded most often in the fishermen's log books."

I stare mutely at a large, pale orange egg sac with a stringy bit of pink attached to it, that has just been cut out from a female cod and is in the process of being weighed. I'm suddenly feeling acute nausea.

"So you mean that trawlers habitually fish in spawning grounds?"

"Of course they do. The fish gather during spawning. There's no money to be earned fishing here at other times."

Now I realise that there is no turning back.

Anders smiles at me with pity and says, "Lying down is usually best. Lie down on the sofa in the mess, it's the best place."

I reach the tiny toilet just before it's too late, cursing myself. Why didn't I take the anti-seasickness pill? How stupid could I be?

I'm lying in the mess with an issue of a science magazine prepared so I have something to hide behind if anyone comes in. Everything else is beyond my control now. The boat cannot stop, and even if it could it would still pitch from side to side. It's too late to take a seasickness pill, nothing goes down anymore. All I can do is lie here and feel how every wave grips hold of me, rendering me more and more powerless, more and more detached from my own element. I close my eyes and hope to fall asleep, but images of fish with gaping mouths and wriggling tails continuously surface in my mind, and so do strange comments and fragments of the research reports and fishing terms I have spent the last few months cramming in: demersal fish, economic zones, catch efforts, pre-summer protection, biomass, spring-spawning Rügen herring, discards and undersize fish. I wonder what I've ventured into, and which direction to take. Will I ever come close to understanding the whole picture and, if so, will it be worth it? I remember the aggressiveness of fishermen on television dismissing the scientists as people who did not know what they were talking about. The tone of the debate is harsh, to put it mildly. Former Board of Fisheries Director-General Per Wramner has stated publicly that he has received death threats. James van Reis, a crown prosecutor at Gothenburg District Court who specialises in fishing-related crime, told me half-jokingly that he scarcely dared to show his face in the Gothenburg islands anymore. The journalist Peter Löfgren, who made the one-hour television documentary *The Last Cod*, said half-jokingly that a fisheries debate he participated in in Sweden was one of the worst experiences of his life – and he said that after having spent years reporting from the Middle East.

Helplessly moved to and fro by waves I'm wishing more than anything that somehow I will be proved wrong, that all of my work so far will be made redundant, that people will come to their senses, allowing me to go and do something else. To dedicate my time to writing about something completely harmless, like therapeutic gardens, literature or conflicts at the Swedish Royal Theatre...

Calmed by these thoughts I fall asleep and don't wake until I hear Bosse quietly placing the frying pan on the cooker beside me. The fridge door opens, some butter starts frying. Soon another smell fills the mess.

He is cooking cod.

A few hours later and a seasickness pill has finally done the trick. I find a surprising joy in just sitting on a brown plastic sofa in the dayroom learning everything you need to know about sonar, plotters and ASDICs. The latter instrument was developed for detecting enemy submarines during World War II and is now used to localise fish schools. An ASDIC can "see" many kilometres below the surface. The sonar is on the worktop in front of me: a screen that shows the seabed in various shades of red, yellow and blue, depending on its consistency. It also would show fish schools but today seems to be an off-day. Not a single spot has shown up on the screen for several hours.

"The herring schools are close to the coast at this time of year. That's why the fishermen have applied for exemptions," Anders explains.

No one on board disputes that fishermen find it hard to make ends meet. Boat after boat is being sent for scrap in the traditional fishing port of Smögen; the fleet there is now down to just six vessels. And then there are the cheap shrimps from Norway which undercut Swedish prices, meaning that fishermen cannot always sell their catches. Fifty-five tonnes of Swedish cooked shrimps were sent to landfill or ground down for fishmeal in the first six months of 2004 because of that. EU subsidies guarantee a minimum price of four euros per kilogram though, which means that fishermen get paid for what they cannot sell and are never forced to slash their prices. Shrimps used to fetch more than eight euros per kilo.

"The glut of shrimps on the market is primarily because there are simply so many of them in the sea," Anders says. "Stocks of other species explode when cod disappear."

Another problem for fishermen is herring – especially the Baltic Sea variety, which is banned from export to EU countries because it contains high dioxin levels. "There were some exports to New Zealand by air for some time," Anders tells me. "The fish was used as feed for farmed tuna but when they heard about the dioxin they didn't want it anymore. Then there were high hopes that Russia and the Baltic states would want the herring but not much seems to be happening on that front. There was a new processing factory built in Västervik with EU money to produce herring for human consumption, but apparently there wasn't any market after all."

We talk about the Baltic cod, which consumers tried to save by boycotting it – a totally useless response, according to Anders. It simply results in the fish, which are caught anyway, being sent to landfill or sold abroad, with the main market being Italy.

"And now Findus are marketing deep-frozen hoki fillets that are transported halfway round the world from New Zealand as an eco-friendly alternative," he snorts.

I almost can't help myself from laughing: this is just too much! But as I look around me at the others cynically smiling while shaking their heads and shrugging their shoulders I realise that when this is over I'll probably be just as cynical as them. What else can one become?

"At least I'm not feeling seasick anymore," I declare, trying to find something just a little positive to say.

The others smile politely. I'm not the first person to feel sick on board this boat, they tell me. Most people make the same mistake and think they will manage without a pill until it is too late.

"We had some foreign marine biologists on board once who almost took offence when we told them to take a pill. Remember the Italian one with ginger hair? He nearly turned green. They have another kind of waves in the Mediterranean; here there is a different rolling to them..."

I ask them who else has visited the *Ancylus* – the Board's Director-General and other top managers, ministers, civil servants? They all fall silent in serious contemplation over the question, and I gather that such visits are not regular occurrences. Suddenly Hans Hallbäck, who is an old-timer, recalls something: there was one agriculture minister in the late 1980s, Mats Hellström, who once came on board. "He actually came out to sea with us. Though he spent most of his time nodding off down in the mess. Every time someone said something he went 'Yes, yes, very interesting' and then fell asleep again."

We are hauling in the final catch at Fladen – a fish bank famous through Swedish radio weather forecasts – when something happens. The most exciting fish caught so far are a skate and a small wolffish with razor-sharp white teeth which both made Anders very enthusiastic. But this is even more exciting. On top of all the other fish in the newly hauled trawl net lies a suspiciously plump-looking female cod of more than 70 centimetres. Anders picks her up immediately and presses her rear gently. Sure enough, a transparent liquid emerges.

"She's ready to spawn," he says quickly. "We'll put her back in."

First he puts her on the scales. She weighs nearly six kilograms. I look anxiously to make sure there isn't any sign of her stomach coming out of her mouth. Fortunately she looks fine. Anders carries her in his arms to the gunwale and throws her in with her head first. He checks for a few seconds that she seems OK, nods with approval when he sees that she does not lie belly up in the water, and then he leaves.

I linger by the rail a little longer. Below I see the female cod still lying in the water with her head above the surface, while *Ancylus* slowly moves away. Had I not read fishermen's stories of fish behaving like this after release, I would almost not have dared say what I am now going to say, but I will say it anyway: she is looking at me. She really is. Perhaps not looking directly at me, but certainly at the boat. Her eyes all round and black, the mouth wide open. And actually, eyelids or not – she looks surprised. For a long time she lies there on top of the waves, slowly moving her tail from side to side, as if trying to establish how she feels. Or perhaps attempting to understand what she has just been through. Then she gently slips beneath the surface, takes her direction and is gone.

THE TRAGEDY OF THE COMMONS

First of all: congratulations! You are the owner of fishing grounds! Fisheries policies actually could be based on this simple but revolutionary fact. The fish belong to you. And to me. And to my children. And to my neighbours' children. The fish belong to everyone; mussels and prawns belong to everyone; starfish, plankton, whales and porpoises are everyone's – the sea is the property of everyone – or, if you will – no one.

With that said, again: congratulations!

There are of course exceptions to this basic rule, some private fishing grounds exist in the world. But by and large the sea is open to all-comers and has been so since the dawn of time; free access is the cornerstone of fishing cultures worldwide.

Unfortunately, the interpretation of this basic principle might not be to *your* advantage though. Up until now it seems, no one has taken into consideration that people who are *not* utilising the sea's resources should also have a say in how they are managed. The principle is instead that everyone who *wants* to exploit the sea should be free to do so. This has been the very foundation of every fishery, a grand liberal idea inherited from the times of the seafarers and discoverers, from the times when the planet was perceived as endless and the oceans were believed to be forever inexhaustible.

But while on land we have gradually relinquished the notion that wild nature belongs to no one; while frontiersmen, pioneers, gold prospectors and colonialists have all staked their claim to virgin land and natural resources, with growing environmental awareness therefore, we have set aside remnants of wilderness as national parks in order to protect nature – on the seas however, our concept of the sea as everyone's – or no one's – has in principle remained unchanged.

Herein lies what is usually referred to as the "tragedy of the commons".

It was Garrett Hardin, professor of human ecology at the University of California, Santa Barbara, who first coined this concept in a 1968 essay on overpopulation, in *Science* magazine. Hardin's line of reasoning has ever since often been used as a model to explain why fishermen time and again have pushed fish stocks below the point of collapse, and routinely – despite depending on fish for their own livelihoods – have raced to deplete stocks far below the optimum level of fish "biomass" that would generate maximum gain for all.

Basically what Hardin was describing was a lack of cooperation in exploiting a freely available but limited resource. He asked readers to picture a village with a common piece of grassland, open to all. This rich and free pasture benefited everyone and was problem-free until the day one villager realised he could earn more by adding another animal to his herd. This villager calculated that while he personally benefited from each extra animal he added to the pasture, the other villagers would share the negative cost of reduced grazing room for their animals. The cost to the first villager was thus negligible and, in addition, he reasoned that if *he* didn't add an extra animal then someone else would – and instead of increasing his gain he would have to share the negative cost with the others.

If just one villager reaches this conclusion, Hardin argues, it would not take long before not a single blade of grass remained, and the tragedy of the commons is a fact.

More than forty years after Hardin penned his famous essay, most of the world's fisheries are nothing but a shining example of how his theory works in practice. There are exceptions of course (as described by 2009 Economic Nobel Prize winner Elinor Ostrom), examples of local communities that succeed in collaborating on fish resources, but as a whole Hardin's thesis is still the most relevant to explain the miserable state of world fisheries. Fishing industry representatives are forever claiming that: "if we don't do it then someone else will". When the moratorium on cod fishing was mooted in Sweden, opponents argued that if the Swedes wouldn't take their share then the Balts, Russians, Finns, Poles and Danes would. When the European Commission proposed a ban on silver eel fishing, Sweden refused, claiming it pointless and unfair unless other countries also banned glass eel fishing. In this context the biggest losers are of course fish from stocks that don't respect national borders, in particular intercontinental migrants like tuna or swordfish. The North Americans, South Americans, Europeans and Asians are all excelling in arguing that there is no point in *them* cutting their catches if other nations don't. It's rather telling that after more than three decades of cooperation, management and joint research programmes on migratory tuna and swordfish under the umbrella of the International Commission for the Conservation of Atlantic Tuna (ICCAT), the organisation has been so stymied by compromises and shockingly inflated quotas that environmentalists now spell out the acronym ICCAT as "International Conspiracy to Catch All the Tuna".

But Garrett Hardin's theory also applies within national borders, right down to individual levels where the same depressing arguments are constantly heard: if I can't fish in cod spawning grounds then somebody else will; if I don't apply for an EU subsidy for a more effective trawler then somebody else will; if I don't take juvenile fish that haven't bred yet, then somebody else will; and if I don't claim my bit of the EU's share in fisheries in developing countries (more of this in the chapter "The Fisheries Agreements"), then somebody else will.

And again and again the outcome is inevitably the same: that those countries and individuals that are the *least* willing to conserve stocks and think of the future are those that set the agenda, while those who voluntarily abstain from overfishing do not gain anything other than having to witness others reaping the rewards of their sacrifice.

That oceans were inexhaustible, whatever man did, was an accepted "truth" well into the 20th century. Actually it seemed very plausible: wild fish are one of the most abundant resources on the planet, they can be harvested in extraordinary amounts with no need for farming or management; and unlike oil or minerals they are forever renewable! At the fisheries exhibition in London in 1883, British zoologist and philosopher Thomas Henry Huxley (a leading 19th century fisheries expert) declared, in response to concerns over the effect that new-fangled inno-vations like otter board trawling and steam boats would have on fish populations: "I believe that the cod fishery ... and probably all the great sea fisheries are inexhaustible; that is to say, that nothing we do seriously affects the number of fish."

Huxley, also the founder of agnosticism, was convinced that no real knowledge in God was possible; still he was convinced he was on the safe side with cod. You might understand why. Back in those days, the fish were so plentiful off the coast of Newfoundland and New England that people joked that they could soon walk across the Atlantic on cod backs without getting theirs shoes wet. John Cabot, who had discovered Canada 400 years earlier, wrote home the now much quoted words that the sea was "swarming with fish which can be taken not only with the net, but in baskets!"

Huxley believed that economic laws made it impossible to exhaust fish stocks, claiming it would be too difficult and expensive to scrape up the last remaining fish. "Any tendency to overfishing will meet its natural check ... This check will always come into operation long before anything like permanent exhaustion has occurred."

But the visionary Huxley, who *could* perhaps have envisioned the effectiveness of modern day factory fishing with trawl nets with openings as wide as twelve jumbo jets and fish-spotting from the air – he could not foresee some other remarkable innovations that would come into play in the 20th century: public subsidies, artificial market mechanisms and EU regulations. Imagining a fisherman who would think nothing of using 500 litres of government-subsidised fuel to catch a few hundred kilograms of shrimp, dump them at a rubbish dump, fill in a form and get paid for the trouble by the EU guaranteed minimum price fund, no doubt requires an extraordinary amount of fantasy. In fact, more of a task for his grandson, writer Aldous Huxley, author of the classic dystopian, science critical novel *Brave New World*.

Who owns the fish? And what is it really worth?

These two questions are never far away at a one-day conference with the intriguing title "Salmon Trout without Adipose Fins – Headless Fishery Management?"

I had never heard of the term "adipose fin clipping" before this conference. I didn't even have the faintest idea that fish had adipose fins, let alone that I one day would find myself attending a conference about them. But here I am, sitting in a large room at Tekniska Nämndhuset in Stockholm, waiting eagerly to hear why anyone would want to remove them.

Seminars about "adipose fin clipping", like most fish conferences, are full of middle-aged men with weather-beaten faces, khaki vests and woolly jumpers. Though to be honest, there are a good few women too, most of whom are a lot younger and seem to be genetics students. They are all knowledgeable, attentive and outspoken, and I soon understand that all belong to one out of three specific categories: anglers, geneticists and fisheries managers. The latter come mainly from the Swedish Board of Fisheries and their sallow features and suits make them stand out from the rest. Only one fisherman has made the journey to the capital, but he seems less interested in adipose fins than in society's indifference to the state of the oceans. "People kick up more of a fuss if they see someone shooting a bloody pigeon than if we kill every fish in the oceans."

Every time he speaks up he is greeted by an indulgent silence.

The geneticists are here for a very specific reason: to voice their disquiet about the extensive releases of farmed sea trout that have been going on in Stockholm since 1957. Adding farmed fish to the gene pool threatens the future of the wild sea trout – a smaller, rounder and less migratory cousin of the salmon. The species' plight is even worse than the wild salmon, according to professor of genetics Nils Ryman who is sitting in the audience.

"At least we know there are wild populations of salmon still out there, but we don't know if that's the case for the sea trout. The situation is very, very serious."

The main arguments levelled against releasing farmed fish are that they are more prone to illnesses and parasites, less intelligent than their wild counterparts, because natural selection has not filtered out the weakest individuals – and then there is the risk of in-breeding. On the positive side, released farmed fish are considered to enhance the quality of life of some half a million people in the Stockholm area who are, to different extents, interested in fishing.

County fisheries adviser Henrik C. Andersson, a thin middle-aged man with round greyish glasses and a smart black suit, takes the podium. He is in charge of fisheries in the Stockholm region, but his agenda is not mainly about commercial fishing. Only around 40 licensed professional fishermen still remain from the hundreds there once were. Since cod disappeared back in the 1980s, herring became leaner and less profitable than they used to be, and pike and perch populations have collapsed; there is not much left for the fishing industry. Andersson's main focus instead is angling – but not of naturally occurring fish, but rather of intro-duced fish such as sea trout and salmon. Around 2.5 million salmon and sea trout smolts are released into Swedish waters every year in restocking programmes paid for by taxpayers and the hydropower industry.

Andersson is talking with a hint of a monotonous northern Swedish dialect, with typical thick "l" consonants when he responds to Ryman's concerns, saying the salmon found in Stockholm's waterways are unique; epitomising the cleanest waters of any capital city and carrying a symbolic value that cannot be measured in financial terms.

"No politician would suggest stopping the restocking programme; it would be political suicide," Andersson asserts.

Then the fact that Stockholm's salmon originate from farmed fish and do not breed naturally, well, "it is simply a reality we have to accept," he states.

"To put it in philosophical terms," he says, "My view is that nature should be there for us humans, so to speak. My job is first and foremost to create a good angling environment and outdoor life for the citizens of Stockholm."

The divergence in attitudes soon becomes clear: are fish there for the benefit of human beings, or are they there for their own sake? And who can claim ownership of the fish?

What are being alluded to are really the absolute crunch questions in fisheries policy. Who owns the fish? How do we best exploit this precious resource, and in what terms do we value it today? Are the 50 million euros of "added value" that the fishing industry generates each year in Sweden the best way to optimise this financial, nutritional and cultural resource? Is it presently being used in the best possible way for "us humans?"

Many would claim that this is not the case: for instance, the Swedish anglers, of whom around 50,000 are active members of the National Anglers' Association. A survey by government agencies Statistics Sweden and the Board of Fisheries revealed that more than 3 million Swedes are actively interested in fishing, with more than 400,000 naming fishing as their main hobby. According to Statistics Sweden, combined spending on fishing in terms of equipment, travel, boat costs, permits, accommodation and other expenses totalled a massive 300 million euros. (This figure excludes foreign anglers who visit Sweden.) In another survey, the Nordic Council of Ministers found that people would spend even more money on fishing if they got more out of it, estimating this latent potential "willingness to pay" at 100 million euros. The "recreational value" of well-being and reduced stress that spending time in nature and reeling in a salmon or two promotes has not yet been estimated in financial terms; but numerous research studies confirm there are major health benefits.

According to Statistics Sweden and the Board of Fisheries, Swedish anglers catch 58,000 tonnes of fish per year, of which half are caught at sea and 11,200 tonnes put back under the increasingly popular "catch and release" system.

A government bill from 2003 summed up all the money spent by anglers on fishing and calculated a turnover of 225 euros per kilogram of caught fish. The value of fish caught by professional fishermen on the other hand averaged 0.4 euros per kilogram.

So how do we value a "free" resource such as fish? At the adipose fin conference I am introduced to a new concept – "the travel cost method", a means often used by environmental economists to try to put price tags in euros and dollars on immeasurable things such as sunsets, rare butterflies and the glimpse of a dolphin. The method consists of asking people how much time and money they would be prepared to spend on travelling to a place offering such an "environmental quality". Environmental economists Tore Söderkvist and Åsa Soutukorva, of the Beijer Institute of Ecological Economics, present a new study at the conference, where they attempt to calculate the economic value of perch caught by anglers in the Stockholm region. By using the travel cost method they have come up with the mind-boggling conclusion that fishermen in Stockholm were collectively willing to pay 1.9 million euros per year to catch an extra 0.2 kilograms of perch per hour.

This is a figure to contemplate at length. Mostly because it illustrates how fundamentally absurd it is to try and put a price on nature. For instance, the economists failed to ask what anglers would be willing to pay to *leave* the perch in the water – in other words, what the fish was "worth" simply by being there. In any case, the calculation remains an over-kill because it is clearly not rational economic reasons that are driving fishing policies today, at least not in national economic terms. All the figures I have seen show instead that economic forces can not be the key consideration in modern fishery policy – but rather political, social or cultural considerations; immeasurable values such as preservation of coastal communities and traditions, and a heartfelt sympathy for the close-to-nature lifestyle that the fishermen represent.

But sadly enough this lifestyle has been undermined exactly because of the good intentions of politicians. Excessively generous subsidies have resulted in too many, too effective trawlers chasing dwindling fish stocks. As a result, the fishing industry is now under threat, as is the fish itself, as well as consumers' access to healthy and affordable seafood.

Maybe the tragedy of the commons is also the tragedy of good intentions. After all, who among us says 'no' when government showers *us* with subsidies? Who among us says 'no' when everything is just free to take?

January 15, 1998, 22.10 pm.
At a studio in Gothenburg the guests have gathered for the *Svar Direkt ("Answer directly")* television debate show. The producers have assembled an impressive number of angry fishermen to fill the semi-circular gallery of studio seats, and it's all prepared for a debate on an explosive political subject. The Ministry of Finance has just published its first-ever analysis of how much the fishing industry contributes to GDP, and it has also reviewed fishermen's incomes – and the figures it has come up with certainly don't look good. The report, very provocatively entitled *Fish and Fraud – on Objectives, Powers and Influence in Fisheries Policy* is claiming, according to the *Svar Direkt* trailers that have been broadcast ahead of the show, that "all fishermen cheat".

There are only two women in the audience: economist Ylva Hasselberg and tax inspector Carin Nicander-Olsson. Among the other faces is a man with hair combed forward, glasses, a suit and a strikingly unhappy expression – the Board of Fisheries Director-General Per Wramner. Behind him, wearing a taciturn look and with grey hair is Social Democratic Party fisheries spokesman Kaj Larsson. Sitting alongside him, the moustached figure of Hugo Andersson, a former Centre Party MP and deputy chairman of the fishermen's trade union, SFR.

In the small ring in front of the audience, together with the TV host, sits economics professor Lars Hultkrantz, a thin man with brown hair and round spectacles. He is the author of the *Fish and Fraud* report – a title, to say the least, unusually outspoken for coming from a government institution. The word "fraud" on an official document has understandably enough aroused strong emotions among fishermen in the country. But now at last everything shall be vented, discussed and broadcast live in front of the Swedish people.

Television host Siewert Öholm sounds the starting gun for the debate by repeating to viewers that the Ministry of Finance has accused fishermen of "only cheating and living on benefits".

The immediate reactions from the fishermen are just as colourful and genuine as in every TV producer's dream: "This is libel," one yells, "defamation," another. "Scandal," argues a third and outright "racism," a forth.

When Lars Hultkrantz finally gets a chance to speak, his dry, academic-sounding Stockholm accent is in marked contrast to the fishermen's broad west coast dialect. Sounding almost like a teacher kindly pleading his pupils to listen, he says: "This is a taboo subject that no one dares talk about. Guess why? Well listen to yourselves now. What we focus on in the report is that this is a huge resource management problem and that it's been there for 40 years and hasn't been solved ..."

"Who says?" a fisherman interrupts.

"I do," Hultkrantz replies.

A peal of laughter from the audience reveals the fishermen's contempt for national economist Hultkrantz's word, and since the catastrophic decline in cod numbers is still not publically known, it is not surprising that presenter Siewert Öholm misses the chance to ask further questions about "the resource", that is – fish in fisheries bureaucrat language.

Instead, this night the studio debate centres solely on economic implications of the report that few audience members seem to have read more than the front cover of. The title was probably not very fortunate, as it diverts the debate from the real objective of the study: investigating how much public money actually goes into the fishing industry and what the taxpayer gets in return. The report also includes interesting research by economic historian Ylva Hasselberg looking at power structures in the fisheries policy area, by using the so-called snowball method – a method to identify key players in any particular sector. Hasselberg asked Board of Fisheries Director-General Per Wramner to name the ten individuals he considered most influential in the field, and then she approached them, asking them to name ten people, carrying on until no further new names emerged. The method yielded 37 names, and the investigation offered a number of interesting insights. Most respondents, for example, could not think of ten names to name, and many of the 37 were named only by one person, meaning for instance that representatives of environmental groups were named only twice out of almost 400 chances.

Sitting in the television studio in Gothenburg are at least four or five people whose names must have been on the snowball list, and Lars Hultkrantz is clearly not one of them. The audience's reaction is immensely mistrusting every time he speaks, even when he focuses on issues that he, as a professor of economics, would be considered to have a certain competence in. Like how much the fishing industry contributes to GDP; how fishermen's incomes are about one third of the national average, and how it is strange that 40 per cent of fishermen officially earn less than 500 euros per month. The fishermen are clearly not used to hearing such statistics from the authorities; still, they show less interest in questioning the figures than the legitimacy of Hultkrantz himself.

"So, how did you get to be a professor?" someone asks. "He's like that other guy, Åke Hallman, who wrote that other report. Get rid of them all; these reviews are costing taxpayers' money!"

"So the fishermen know more than the researchers?" asks presenter Siewert Öholm rhetorically – to vehement affirmation from the audience.

The fact that a large government agency like the Board of Fisheries does not like to question the industry over which it presides becomes clear when Per Wramner, the Board's Director-General, grabs the limelight on live television and presents a completely new figure. The Hultkrantz report had put the cost to taxpayers of fishery management and industry support at 42 million euros per year – almost half of the value of landings. Wramner, looking very uneasy, says this figure is totally wrong and in reality is just 24 million.

Responding to this bombshell, an astonished Hultkrantz asks how such a massive discrepancy could arise. But the TV host interrupts, complaining that the debate is becoming bogged down in "details". Neither Hultkrantz nor the audience find out how the 18 million disappeared.

Öholm moves on, asking MP Kaj Larsson to explain the extent of his responsibility for the report – since now the implication is already that the report is a scandal, libellous and riddled with errors. Larsson denies all responsibility, declaring he has nothing to apologise for and stressing that he is also critical of the report, especially its accusation of financial chicanery among fishermen.

The fishermen in the audience cheer and one shouts out: "He's a sensible person, let *him* investigate next time!" (An idea that was fulfilled five years later when Larsson resigned as an MP to become chairman of the fishermen's trade union, SFR's, special cod group.)

Researcher Ylva Hasselberg is allowed a few moments of airtime, during which she argues that the industry is overcapitalised and therefore tends to overfish even when stocks are too low for fishermen to pay their loans and mortgages.

The debate is wrapping up when tax inspector Carin Nicander-Olsson chips in to say that the National Tax Agency is only too happy to help small business owners fill in their tax returns which she understands are too complicated for them to fill in.

A chastened and almost stammering Hultkrantz makes a final attempt to assuage the fishermen, stressing that the report does not aim to label them as cheats but to point out that politicians must shoulder responsibility for ensuring that commercial fishing interests are not allowed to ride roughshod over consumers, sports fishermen and environmental groups. But TV host Siewert Öholm does not seem interested in these arguments. Instead he holds the report with the reprehensible name aloft and asks the audience:

"So shall we throw this one in the bin, then?"

The audience agrees.

The "miscalculated" 18 million had become a public truth.

Adipose fin clipping, is, believe it or not, a useful tool in the conservation of wild salmon and its cousin the sea trout. Much of the Baltic populations of these two species consist of farmed and released fish that are indistinguishable from their wild cousins, but now this is going to change. Since 2003, the 2.5 million or so salmon and sea trout smolts that are released into the sea and major lakes like Vänern, Vättern, Mälaren, Hjälmaren and Storsjön must be marked by removing the adipose fin – a small fin between the dorsal fin and tail – to facilitate identification, and as a tool in a future ban on the fishing of wild fish.

At a cost of 0.15 – 0.20 euros per removal, adipose fin clipping does not come cheap. In the paradoxical world of legislation, it is required under animal protection laws that the fish must be given a painkiller before the fin is clipped and that only people approved and trained by the Swedish Ministry of Agriculture are allowed to perform the procedure.

Still, the geneticists at the adipose fin conference are interested in debating the ethical aspects of this procedure.

"What happens," asks professor Nils Ryman, "when animal rights organisations find out that millions of fish are being mutilated? Imagine if militant vegans try to liberate the smolts!"

He is half joking, but the absurdity of administering anaesthetics to fish that are likely to end their days a few years later choking in a net or slowly dying with a hook in their mouth hangs heavily in the air, although nobody mentions it.

I take the chance to ask a quite fundamental question that has been bothering me ever since I first saw the term "adipose fin" in a Board of Fisheries press release.

"What does the adipose fin actually do? Does it have any specific function?"

The room falls silent for a couple of seconds. Necks turn to see who has posed such an odd question. County fisheries adviser Henrik Andersson pulls himself together on the podium and shrugs.

"Nothing much."

Then he corrects himself, adding quickly:

"As far as we know."

I am taken aback, though I shouldn't be.

"We don't know?"

Professor Ryman in the audience has mercy on me, turns around, and explains.

"Well, you know. It does make it sexier."

One mystery remains: why allow salmon fishing in the open sea at all? Since there you have the likelihood of catching wild salmon alongside introduced fish. Instead you could only allow fishing on the river estuaries where you can target the released salmon that return to spawn, and you wouldn't need adipose fin clipping at all. Catching salmon at sea, I note to a Board of Fisheries official afterwards, means threatening the wild stocks since the selection only comes after the fish has been caught! How does the Board expect wild salmon with unclipped adipose fins to survive the longlines?

"Survival rates are higher with longlining than in nets," he tells me. "You unhook them and put them back. Fishermen have no problem putting fish back; they already do."

"But," I ask, "Why not just ban fishing of salmon at sea?"

"It's no big deal," he answers. "Salmon fishing at sea will be phased out anyway and replaced by coastal fishing with traps; that's what the ban on drift nets will lead to."

I am confused, but the official has to rush to catch a train and leaves my follow-up question unanswered. "No big deal" – what did he mean by that? It certainly seemed like a big deal when the Board of Fisheries earlier had warned the EU that banning fishermen from using 21-kilometre drift nets might wipe out the entire Baltic fishery!

The explanation to the fervent opposition on a ban on drift nets only comes to me later. WWF fisheries expert Inger Näslund tells me that it was the Danes and southern Swedes that were the main source of opposition. For them it wasn't valid to say that you might as well catch the salmon in their native rivers; they wanted to keep the fishery in the open sea in the southern Baltic.

"It's obvious why. Think about it. How many salmon rivers do they have in Denmark? None."

On returning home from the conference I happen to open *Song for the Blue Ocean*, a book by the American ecologist Carl Safina that had been lying on my bedside table for some time. Safina starts by describing the plight of different tuna species that are being overfished worldwide because of their transnational migration routes, high value, and because every nation at every moment is trying to get as large a share of the catches as it possibly can. Then he moves on to the legendary rich salmon waters of the north-west American seaboard. On visiting the area in 1996 he found that fishermen's incomes and catches of all salmon species had plunged by 97 per cent in a decade. But here, over-fishing is only part of the cause: massive clear-cutting of hectare after hectare of riverside forest has adversely affected the smolts, depriving them of shade and raising river temperatures. To make matters worse, harmful sediment had started to run into the rivers when there are no root systems to prevent riverbank erosion. Safina describes what he sees from an aeroplane window: square mile after square mile of bare, clear-cut, stumpy ground, and curtains of forest alongside the roads and coast – left by the timber companies to mask the full extent of the logging from the public eye.

Before the arrival of Europeans, Safina tells us, these areas were home to three First Nation tribes – the Salish, Tlingit and Kwakiutl – each with its own rich culture. The surrounding forest met all their needs in terms of food, fuel and protection; it gave them a good life. In addition, the tribes benefited from an annual miraculous event – the migration of the "Swimmer", the fat, silvery salmon that voluntarily came swimming upstream in all the rivers. All three tribes regarded salmon as sacred. The Salish, Tlingit and Kwakiutl thought the various salmon species – coho, sockeye, pink, chum and chinook – were different tribes of the salmon people, whom they believed lived in a magical city out in the ocean, on the horizon. The heroic upstream migration, climbing mountains, even leaping waterfalls, the Indians believed was a sacrifice of the salmon people to humans – a peace offering that should be accorded utmost respect. The tribes all had ceremonies that venerated "the Swimmer", the magical friend who could vanish into the endless ocean as a small smolt and return years later a hundred times bigger, fight its way up the river, into the mountains, even rising above cloud level, to eventually spawn – on its own deathbed.

But for me what is most striking about Safina's account is not the story of the tribes' reverence for the salmon – but the fact that the salmon's existence and the rich forests in which people lived, gave them so much free time on their hands. The evidence for this is, among other things, that all their possessions, every simple object, from spoons to fish hooks, were decorated with ornate designs to such an extent that we today would call them works of art.

I put the book down and reflect on man's impressive potential, expressed in so many different ways in different times. But was it really cultural progress when we stopped decorating fish hooks and began manufacturing ten-mile-long drift nets? Was it really progress when we stopped calling the salmon our friend, and started estimating its value purely in euros and cents, even thinking it was completely rational and sensible to use the "travel cost method" to prove that it has a value at all?

Professor Lars Hultkrantz recalls the night in the Gothenburg television studio as a set-up. Two days earlier, a journalist had called to invite him to take part in the debate show, informing him that guests would include eight representatives from the SFR fishermen's association and Per Wramner, Director-General of the Board of Fisheries. Hultkrantz accepted the invitation, although he had a bad feeling about it.

"The person who called me up was a young journalist who clearly hadn't read the report. I got the impression that no one else on the editorial team had either. Instead it seemed like the SFR had provided all the questions to be asked."

Minutes before the live broadcast, as presenter Siewert Öholm flicked through his papers and studio staff were testing microphones, asking the audience to sit as closely together as possible, Hultkrantz tried to greet

Wramner. The two had been colleagues at the Swedish University of Agricultural Sciences, where both had held professorships.

"But he was very evasive and clearly nervous," Hultkrantz recalls. "And I was about to find out why."

The background of the *Fish and Fraud* report had actually to do with both Wramner and Hultkrantz. Back in 1994, it was Wramner who had commissioned Hultkrantz to write a report for the government on sport fishing in Sweden.

"To me it was a totally innocent little report on the completely uncontroversial topic of how to promote fishing tourism in Sweden."

But Hultkrantz was soon to find out the topic was neither innocent nor uncontroversial at all. When asked to present his interim findings to Wramner's reference group, he found Reine J. Johansson, the charismatic chairman of the Swedish fishermen organisation SFR, in the room. Johansson, a former leader of the Young Social Democrats and Swedish Municipal Workers' Union official, had been the most influential figure in the Swedish fishing industry for more than a decade. Hultkrantz's jaw almost dropped open when he heard what Johansson had to say about his report on Swedish sports fishing.

"He said, and I quote, '*We* have decided not to publish it.'"

Hultkrantz never found out who "we" were or why "we" had decided to pull the report.

"Before the meeting, there was no doubt whatsoever that the report should be published. Wramner had been happy about it. When he rang me afterwards he was very apologetic, suggesting having the report published by someone else, which is what eventually happened."

The study was published by Umeå University and later incorporated in a government report on sport fishing. And soon afterwards Wramner demonstrated that at least he wasn't unhappy with Hultkrantz's work, by contacting him with a new proposal. Newly appointed Director-General of the Board of Fisheries, Wramner wondered if Hultkrantz had any students who might be interested in researching salmon fishing in the Baltic Sea and investigating its importance to the fishing industry.

"I had a student who had started reading national economics after a successful business career, and he travelled round the Baltic documenting fishermen's homes and cars – comparing findings with official figures that indicate fishermen barely have any income at all. He also found out about some unreported landings on the island of Bornholm and found that ex-fishermen were awarded interim fishing licences and spent their summers salmon fishing, and that the catches were not reported in the official statistics."

The report was published under the rather unexciting title *Socioeconmic Studies of Salmon Fishing*, by Björn Finn and Johan Snellman (Centre for Research on Transportation and Society), and it only took a couple of days before a science journalist from the Swedish public service radio rang Hultkrantz asking for a copy.

"I said, 'Of course', and the journalist said, 'By the way, you know it's classified, don't you?'"

Hultkrantz certainly did not, and was all the more amazed to hear that the reason was "foreign security". When the story hit the media, the secrecy was lifted but Hultkrantz still doesn't know what was behind the classification in the first place.

"The official reason was probably something to do with the negotiations between Baltic Sea countries about salmon fishing at around that time. But in retrospect, we believe it was our discovery of the abuse of interim licences that was the sensitive issue."

When a year later professor Hultkrantz was invited by the Swedish Ministry of Finance to come up with ideas on public sectors that needed to be evaluated from a financial point of view, it was only natural he suggested to do an in-depth study on the economic benefits of the use of fisheries resources. The responsible minister agreed this was a forgotten corner of Swedish national economics that needed to be looked into.

"But at the end of the day it turned out no one really wanted to see what was in that corner."

During Per Wramner's tenure as Director-General of the Board of Fisheries, from 1989 to 1998, Sweden's annual cod catch fell from 50,000 tonnes to 20,000 tonnes. These nine years also coincided with a splurge of subsidies to the fishing industry in the wake of Sweden joining the European Union in 1995 – at a time when Swedish and European politicians could easily have learned from the collapse of stocks in Canada and applied the brakes. I meet Wramner, now with grey hair, at Södertörn University near Stockholm, where he has returned to academia as professor and research group leader at the Coastal Management Research Centre (COMREC). The campus here is something of a haven for young, ambitious academics and innovative institutions. There is a buzz about the place and students mill around under the high wooden ceiling in the entrance hall. Wramner buys me a cup of coffee from the dispenser in the staff canteen and confides that he is happy he is no longer restricted in what he can say. He has been waiting, he says, for this opportunity to speak out.

"It's been frustrating for me, but I can speak openly now. As director-general you have to be loyal to your employer, that is the government, and I was."

A trained biologist, professor, who had previously been head of a Swedish environmental NGO, and currently on the board of WWF, Wramner is not the usual former head of a government agency. And indeed he has mixed feelings about his time at the Board of Fisheries.

"There was a lot of internal friction between biologists and economists," he recalls. "The fishing industry was very influential. All the agriculture ministers during my time – Karl-Erik Olsson, Margaretha Winberg and Annika Åhnberg – had close links with the fishermen's association and all three of them were disastrous for Swedish fisheries."

Wramner starts off by explaining how impartial scientific findings gradually get turned into wishful thinking that suit the politicians better. His description chimes with the experiences of Linköping University student Johanna Eriksson as recounted in her thesis *Preparing a Cod for the Negotiation Table – From Science to Politics*, and Jan Thulin, director of the Institute of Marine Research (1994 – 2000), who once referred to his own role as being "first mitigator" of scientific results. On a piece of paper he sketched the four stages with arrows between them: "research – negotiations – horse-trading – environmental carnage".

Wramner talks me painstakingly through the process that has led to the carnage of cod in Swedish waters. First, the Institute of Marine Research (which is run by the government authority the Swedish Board of Fisheries) conducts trawl surveys to estimate stocks, then does random controls of catches, and checks fishing boat logs.

"Already at the log book stage the first element of uncertainty arises.

It's a well known fact that records are falsified, and if you trust those figures you have a problem."

The figures are then "processed" by the Institute of Marine Research, after which they pass on their findings to the International Council for the Exploration of the Sea in Copenhagen. But, as Johanna Eriksson writes in her thesis, the problem is that "the research institute is connected to the government agency" – in other words that a government agency whose sole purpose is to regulate a certain industry, might be reluctant to allow its research unit to produce data supporting a closure of that very industry. This is what Jan Thulin means when talking about "mitigation" of the data.

In Copenhagen, Swedish researchers meet their foreign counterparts, who themselves may have been influenced by internal processes in their home countries. The results are then assembled and taken care of by the ICES Advisory Committee on Fishery Management (ACFM), known for its cautious approach and well documented propensity for penning impenetrable documents, virtually unintelligible to a layperson. Somewhere in these documents, the magic figures for recommended quotas for the coming year usually appear. These recommendations are then sent to the Board of Fisheries' marine resources department, which then submits an expert opinion to the Swedish Ministry of Agriculture. The opinion is based on the scientists' reports, but also on industry opinions – another facet of how the figures are "mitigated" and amended to reflect the socio-economic dimensions that so often are mentioned in this context. The Ministry of Agriculture then incorporates *political* aspects: relations with other fishing nations and public opinion.

Once Sweden has reached its position, it has to align with other countries in the EU where environmental and consumer groups have even less of a voice. The European Commission has also produced a proposal for a quota, based on analyses of the STECF (Scientific, Technical and Economic Committee on Fisheries) which also incorporates socio-economic factors into its recommendations. The final step is to reach an accord with Russia, Norway, or other neighbouring states.

Thus there are at least seven different stages at which the hard data submitted by scientists can be "mitigated". Ultimately, the thick wads of documents prepared by the experts are condensed into a political one-pager where socio-economic and political considerations crowd out the scientific data. This is the document that ends up on the fisheries ministers' table, and this is what allows the management of wild natural resources to become environmental carnage.

"This is the formal explanation," says Per Wramner as we are sitting at the coffee table. "And then you have to ask yourself why the ministers have listened so much to the industry and how those links could have become so strong at different levels? Because they have."

Wramner tells me that when the Centre Party's Karl-Olov Olsson served as agriculture minister his party colleague, Hugo Andersson, worked as a fisheries adviser at the ministry. Andersson later became deputy chairman of the Swedish fishermen's association SFR (and after that the chair of the North Sea Advisory Council). Social Democrat MP Kaj Larsson was a government fisheries specialist for many years and was so well established that he was seen as an "unofficial fisheries minister". After leaving parliament, he also joined the Fishermen Association SFR – as chairman of its cod working group.

But particularly one person, Wramner says, exerted heavy personal influence over staff at the Board of Fisheries, and that was former SFR chairman Reine J. Johansson, later to become chair of the Baltic Sea Regional Advisory Council.

"He scared a lot of people. To be blunt, he was an occupational hazard. I've seen people leave their rooms in tears on more than one occasion after Reine turned up, which he usually did unannounced. Some people went on sick leave and one even took early retirement because it affected him so badly. Fishing is a tough industry; I'd say it arouses similar feelings to wolf hunting. Government officials are not used to people using that tone of voice and many couldn't handle it. They were basically scared."

Wramner himself was the victim of several anonymous death threats during his time as Director-General. The intimidation reached its peak when the Board imposed a moratorium on salmon fishing.

"It's interesting that Sweden wasn't able to impose a ban on cod fishing in 2002 because during my time we brought in a unilateral ban on salmon fishing in the Baltic for a few months without even asking the EU. That really put the cat among the pigeons: we offered the fishermen compensation during the ban, based on their declared income, and that wasn't popular at all. I got telephone death threats twice – once at work and once when someone called me at home and said, 'You've been messing us around and we're going to make sure we get a new Director-General, and there's only one way'. The police took the threats seriously, partly because they'd overheard radio traffic between fishermen out at sea and had heard that I was going to be 'marked'."

Wramner was advised to take a course in self-defence. And he had to give plans of his apartment building to the police, and put up with security guards being stationed in his apartment block when he invited Board of Fisheries board members home for a drink. On one occasion the Security Service of the Swedish police advised him not to travel to a meeting on the Baltic island of Gotland. Clearly, all of this must have been an extreme, frightening and frustrating situation for a public servant seeing his main task as serving the government, and yet time and again finding himself let down by that same government.

"Whenever fishermen decided they didn't like our decisions they appealed to the government," Wramner recalls. "The ministers often listened more to the fishermen than they did to us. When the EU decided that large trawlers had to have VMS transmitters to enable the authorities to control them, Reine Johansson lobbied agriculture minister Annika Åhnberg to say no to the proposal. Sweden was the only EU country to object – along with Spain, which has an infamously strong industry lobby. That was a disgrace, in my view. Embarrassing. The government went against its own fisheries agency."

But what about the *Fish and Fraud* report for which professor Lars Hultkrantz had to withstand such a barrage of abuse? How could Per Wramner claim – on live television – that Hultkrantz's calculations were so wide of the mark?

In summary: according to Hultkrantz the fishing industry "cost" taxpayers 42 million euros per year in direct subsidies, benefits, financial support, administration, research and regulatory controls. Hultkrantz calculated this figure to be around 40 per cent of the production value of caught fish. Hultkrantz also calculated an hourly value added for fishing. The average value added across all sectors of Swedish industry was 28 euros per hour in the mid-1990s; for the fishing industry it was 8 euros – far below sectors with low skills requirements like timber processing

(24 euros) and cleaning (13 euros). The data revealed a tripling of government fisheries expenditure since Sweden joined the EU in 1995, and in the same period employment had fallen. Each licensed fisherman "cost" 14,120 euros of taxpayers' money every year, despite their declared average wages being very low. The report also suggested that fishermen had "numerous opportunities to cheat" and "there are various indications to suggest that these opportunities are used on a not-inconsiderable scale". The fact that a quarter of fishermen on Sweden's south-east coast earned just 0.9 euros per hour (including unemployment benefit) from their occupation suggested financial irregularities, the report concluded, suggesting that fishermen either had undeclared incomes or were "living from thin air".

The latter hypothesis was considered "unlikely".

These findings inevitably angered fishermen, provoking claims of libel and persecution. But why had Wramner accused the report of inflated figures for government expenditure? Why had he questioned Hultkrantz's data in the TV show?

"My main criticism was that a major part of the Board's spending goes to things that have to be done anyway, whether we have a fishing industry or not, like doing stock assessments. We need to know how many fish there are in the sea, just as we have to know how many animals there are on the land."

"But," then Wramner adds unexpectedly, "I also felt I had to defend the Board's existence. There were people out there who wanted to close us down and a few years earlier we were forced to cut the organisation by 30 per cent. I saw the figures in the report as part of a campaign to shut us down."

Wramner explains that he deducted costs relating to general conservation, liming, restoration of waterways, the Fisheries Inspectorate, and research conducted on behalf of the public and anglers – expenditure that Hultkrantz estimated at 450,000 euros when they were put to him at a later date.

I find Wramner's maths hard to follow, but I have managed to unearth an interview with him in the *Yrkesfiskaren* SFR trade magazine. The interview appears in an issue of the magazine that is in its entirety dedicated to the Hultkrantz report, stuffed with headlines such as "Libel against hardworking fishermen", "Offending the entire profession", "Mad maths", "Defamation without any proof", "Full of inaccuracies", and "Smearing of honest people". Under the headline "Speculative methods and false claims" Wramner derides Hultkrantz's methodology as befitting a newspaper article rather than a scientific study, saying it is misleading to assign costs for the Institute of Marine Research in Lysekil to the fishing industry because they would still have to be taken "if there wasn't a single fisherman left in the country". Wramner bristles when *Yrkesfiskaren* tells him that a footnote in the report insinuates the Board had failed to file some of its paperwork. "That's an absolute lie!" he told the magazine. "We file everything. It is defamatory and untruthful to claim that we consciously avoid filing documents!"

The tone is shrill, to say the least. And eight years later, professor Per Wramner is drinking lukewarm coffee from a plastic cup, looking uneasy. "A lie", "defamatory"? How could he have allowed himself to use such language? I wonder.

"My press secretary advised me to go for it, and to be very straight-forward to make sure I got the chance to speak in the TV debate, and I did. But maybe I went a bit too far and I regret that. I appreciate Hultkrantz and I thought there were a lot of good things in his report, even if there were parts that I was very critical of. But my priority was to ensure the survival of the Board. I have apologised to Hultkrantz subsequently."

We could of course spend ages on the rights and wrongs of this story. Hultkrantz tells me that he threatened to report Wramner to the Ombudsman of Justice, and that the new agriculture minister Margaretha Winberg finally half apologised to him. The matter was raised in parliament, during which the then minister, Annika Åhnberg, explained to furious MPs that the report was addressed *to* the ministry and had not been commissioned *by* the ministry, a crucial distinction that would take away any blame from the government. The MPs hammered in the word "defamation", and one called Hultkrantz a "joker", another a "comic book author". The report, incidentally, was not followed up until eleven years later (in 2008) when the Swedish National Auditors performed an audit of Swedish fisheries, reaching very similar conclusions.

We talk for a long time, Wramner and I. He tells me about the culture at the Board of Fisheries; how major issues were discussed in "informal" conversations initiated by the fishing industry. He gives me names of people who played a key role in making sure that the politicians listened more to the industry than to the scientists – various officials and MPs of various parties who have acted as the "extended arm" of the industry.

It was these close contacts that led to the non-publication of Hultkrantz's sports fishing report. Some MPs belong to the same evangelical Free Church as the owners of some of the largest fishing vessels on the west coast. One MP is chairman of the Swedish Water Owners' Organisation, who leases fishing waters to fishermen on the east coast. Other MPs have consistently acted in the interests of fishermen, perhaps motivated by ideological considerations. The group that decided that "we" would not publish the Hultkrantz report on the future of sport fishing tourism contained MPs from most political parties.

"I did of course have an ulterior motive in commissioning the sport fishing report, and people understood that. But no one wanted the fact that fish would be worth more if they were fished by anglers than by professional fishermen to be highlighted, so everyone said the report was no good. It really upset me but I couldn't do anything as long as I didn't want to openly obstruct the government."

Wramner insists he had no idea that fish stocks were in the dreadful state they turned out to be. But what was he thinking in 1992 when Canada's cod fishery collapsed?

"I thought it could never happen here. Sure, we had a crisis in 1995 and had to cut quotas, but the real crisis didn't happen until after I left the Board in 1998. Things started going downhill in 1999 and have continued that way ever since."

Wramner applauds the current ecosystem approach and its multi-species management model. "At the start I didn't see the fish feed industry as a problem. If there are fish then it's best to use them. But since then we've learned how herring change the ecosystem and the scale of the disruption that has occurred. The more we learn, the more we understand that we have to be very cautious."

BEFORE THE BIG BOATS CAME

Smögen, Bohuslän, April 2004.

It's impossible to stroll incognito along the quayside at the picturesque seaside town of Smögen in April. The tourist season is still a long way off; the boutiques, ice-cream stands and souvenir shops are all closed, and the empty, narrow wooden jetty lined by boathouses brings some dream-like scenario from a movie to mind. The wind from the sea whines around the small houses; a seagull unexpectedly mewls from a lamp-post.

In the 1970s when I used to come here with my grandparents in summer, the wooden fishing boats would lie two or three deep on the moorings. We bought fresh shrimp from the fishermen and ate them straight from a brown paper bag. Then we swam in the freezing water, amongst the blue and red jellyfish. The jetty at Smögen was a universe all of its own, a mysterious micro-world with clucking sounds, buoys of glass, fishing nets, sheens of oil shimmering on the water, crabs on the bottom, and the unmistakeable aroma of seaweed, tar, fish and hot dogs in the air. Now only six out of a hundred or more fishing boats remain at Smögen, and on this April morning not a single one is to be seen. Of the active fishing community dating back 400 years, there are only two visible signs: the newly refurbished fish auction building at the end of the quay and "Gösta's", the fishmonger's opposite – the only place open on this windy April morning.

Gösta is a man in his fifties with steel-rimmed glasses, protruding ears, and a red cap and shirt. Selling me coffee and a shrimp baguette, it does not take him long to work out who I am.

"She's a reporter," he calls out jovially to everyone who enters the shop. "Talk to her!"

He winks, jokes and acts the tease. And everyone *does* pass through his door sooner or later: the people who work at the auction house, fishermen, former fishermen, summer tourists who have stayed on and moved in, wives, schoolchildren.

"Hey reporter!" he calls out with his west coast dialect. "You should have a chat with him! Reporter, do you want a cup of coffee? Reporter, you should really have a chat with him over there, he's single, you know!"

Gösta has also already twigged the purpose of my visit. I'm here because Smögen happens to be home to a fisherman who has gone against the tide and fought the authorities and his own trade federation in an attempt to create new economically and ecologically sustainable alternatives. Bo Hansson's campaign to persuade the Board of Fisheries to consider selective fishing methods and to "try and bring together widely differing views on fisheries in a constructive way" won him the WWF Mannerfeldt Prize and a cheque for 10,000 euros in 2003. Back in 1996 he had already invented a trawl net with rectangular mesh that halved by-catch of undersize nephrops (also called Norway lobster). This was a kind deed because juvenile nephrops become blinded by daylight when brought to the surface and die when thrown back into the ocean. Bo Hansson is in other words one of few fishermen who have worked with researchers and environmental groups, seeing cooperation as vital to the survival of small-scale coastal fisheries.

"Ah, you're here to meet Bosse from Kornö island," people tell me on hearing this news from Gösta. Most of them shake their heads. "He's all friendly with Maria Wetterstrand and the Green party nowadays, isn't he?"

One man shakes his head particularly energetically:

"That Green Wetterstrand woman ... There should be a bounty on her head."

Bo Hansson arrives wearing a cap and astride a brand new moped. Without unnecessary words, we leave the fishmonger and drive through Smögen, past the yellow Free Church chapel, up the hill by the playground where the children's slide wears a big advertisement for a seafood brand, past numerous small square houses clad in grey asbestos panels standing on bare, windblown rocks – to the street where he lives, on a hill facing the harbour at Hasselösund. He is evidently used to the looks he gets when he picks up "the reporter" at the fishmonger's. Hansson is an intense character, who has always had the ability to think innovatively and, no doubt, ruffle a few feathers. In 1969 he started the Tam Tam discotheque in one of Smögen's former herring warehouses, and in the mid-1990s he co-founded a new small-scale fishermen producer organisation, a breakaway from the national union of fishermen, SFR.

"That was the only way," he says, sitting in his study with a view over the nearby houses and rocky outcrops. "I sent motion after motion to the SFR congress to try and get an agreement on regulating fishing intensity, but they turned them all down."

Bo Hansson – in Smögen never referred to as anything other than "Bo from Kornö island" – is a man in his sixties with a short grey beard and round glasses. He speaks very fast in heavy Bohuslän accent, knows his subject extremely well, and is the sort of person who likes to back up what he is saying by showing you hard evidence on paper. While we are talking he repeatedly gets to his feet to pluck reports from his bookshelves, peppering his conversation with figures and technical terms like "eutrophication" and "demersal species". You can tell undoubtedly that he has worked hard to try and convince the world of what he is about to explain to me now. Finally he spells it out in summary after an hour or so, while I am sitting exhaustedly contemplating the mountain of papers on the table in front of me:

"Instead of many small boats fishing with nets with large mesh sizes, we now have a few large boats fishing with tiny mesh," he explains. "Instead of conserving fish until they've reached a good size, there's a race to fish the juveniles."

His eyebrows rise over the round glasses, and he looks at me with a knowing purse in the corner of his mouth.

"And that's not a very intelligent way of doing things."

Smögen's problems really began in the late 1980s and were largely precipitated by the United Nations Convention on the Law of the Sea (UNCLOS), signed by the international community in 1982. The convention sought to address years of confrontation over fishing rights in international waters – conflicts that sometimes turned ugly, as when Iceland unilaterally extended its exclusive fishing rights from 3 to 200 nautical miles off the Icelandic coast, leading to the 'Cod War' with Britain and copycat actions by several other countries. The convention, signed by 117 nations at Montego Bay, Jamaica, defined territorial waters as the 12 nautical miles closest to a state's coast baseline and exclusive economic zones (EEZ) as being the closest 200 nautical miles. EEZs give coastal nations sole exploitation rights over fishing, oil and gas exploration and other economic activities.

This convention was certainly good news for Sweden's neighbours, but for Sweden, and in particular the Swedish North Sea trawling fleet, it dealt a devastating blow. The large Gothenburg-based vessels that used to fish herring and mackerel far out in the North Sea suddenly found themselves confined by Danish and Norwegian EEZs to a narrow corridor of territorial waters on the Swedish west coast. Fortunately for the small-scale fishermen operating out of ports like Smögen, this event coincided in time with an unusual boom in cod reproduction in the Baltic Sea, so the large trawlers that used to fish in the North Sea simply switched to fishing the plentiful Baltic cod. But in retrospect the Baltic cod boom eventually impacted negatively on Smögen's small-scale fishermen because it encouraged the long distance west coast fleet to keep fishing at a time when it should have been downsizing in the wake of the UNCLOS declaration. Matters were not helped by the overheated 1980s economy either; cheap loans and benevolent government subsidies were easy to come by and the scale of the new investment surpassed anything seen before: the Swedish fleet grew bigger than ever.

"The policy at that time was that fishing should be run as an industry, just like any other business," Bo Hansson says. "And that was a recipe for cultural destruction. North Sea fishing used to be a hunting culture, and a hunting culture does not exterminate what it is living off. But if you turn fishing into an industry then things turn out differently ..."

In Smögen at that time fishermen still thought in the old ways, and ever since 1964 they had agreed a series of voluntary restrictions to protect fish stocks and their own livelihoods. The main rule was that fishermen could only fish on Mondays to Thursdays from 5 a.m. to 5 p.m. Social pressure ensured that the rules were respected and everyone could see there were enough fish for all. Everyone also knew that relaxing the restrictions would create a risk of overfishing.

"But around 1984 we started seeing the big trawlers from Gothenburg coming up here by us to fish at weekends, during the evenings – whenever," Hansson says.

Equipped with the latest technology, including asdic sonar and modern trawl nets capable of trawling rocky seabeds, these big vessels were now not only targeting cod in the Baltic, but could also fish close to shore on the west coast, in waters where in the past only local fishermen with intimate knowledge of the area had been able to fish. They in turn, bound by their voluntary agreement, had to stay in port from Friday to Sunday helplessly having to watch the newcomers plundering their fishing grounds.

"They behave like predators, those boats from the south – whatever they fish," a fisherman on the island of Nord-Koster says in a book entitled *Protect the West Coast – While There's Still Time* (*Värna Västerhavet – medan tid är*). "They figure they might as well keep going for as long as stocks last, saying 'if we don't do it then someone else will.' When the fish run out they just move on. And we're left with all our 'old crap' and our dependence on local fishing ... But is it really better to spend millions on a boat and then have to fish night and day to pay off loans and interest, only to find out that you can't make ends meet anyway? It just leads to overfishing. Just imagine the amount of fish they have to catch to earn the same amount as small boats like ours. They sweep the seas like vacuum cleaners."

Bo Hansson describes how the Swedish fishermen's association failed to resolve the conflict, resulting in Smögen's fishermen abandoning their voluntary agreement on fishing restrictions in 1988 and unleashing a 24/7 free-for-all. When Sweden then joined the European Community in 1995 and subsidies started raining down on a sector whose main problem was that too many boats were chasing too few fish, matters got even worse.

And what has been the result of all this? Hansson points at the piles of documents on his desk. Reports such as "Sea in Balance" by the Swedish Environmental Protection Agency, "Sector Targets for Ecologically Sustainable Development" by the Swedish Board of Fisheries and the *Green Paper on the Reform of the Common Fisheries Policy* by the European Union – along with his own notes and documents; kilos of paper that have not changed anything.

"They fished until everything was gone," Hansson fulminates. "And who are the victims? We, the coastal fishermen. The others just move on and start fishing somewhere else."

He looks out of the window, looking genuinely preoccupied. Norwegians and Germans are buying house after house in the town, he tells me. His own fishing boat he has sold for scrap in return for EU and Swedish government grants. Smögen may have to turn to fishing tourism if it is to have a future, Hansson suggests, and immediately produces two other documents from the shelf on the topic. One is a 1996 government report on sustainable development in Swedish coastal areas and the other is a consultation paper submitted by his own organisation, the Norra Bohuslän Producentorganisation. I leaf through some pages and then look out over the rocks contemplating convoluted sentences like: "The measures ought to be incorporated into socio-economic development through bioeconomic consideration of the actual needs of coastal communities", and suddenly I feel very, very tired.

"Hey, reporter, how's it going, reporter?! You should talk to that bloke over there, he's an old-timer!"

I am back at "Gösta's fish" after interviewing Bo Hansson. Gösta blinks benevolently at me from beneath his baseball cap and nods his head towards a tall, upright man with white hair, who's wearing a checked shirt under his jacket. Sven Rödström, manager of the Smögen fish auction from 1978 to 2002, introduces himself. Polite and mild-mannered, he is happy to sit down with me at one of the white plastic tables inside the shop for an interview.

He starts talking thoughtfully, before I even ask my first question. "I thought things looked very promising altogether in the early 1980s, you see. Even in the early 1990s I thought there was a future, but since we joined the EU things have mostly gone wrong. People are getting paid millions to send small boats to scrap and getting subsidies to build enormous trawlers costing some ten million euros. There's no chance you can make a profit on that sort of investment. Some of the boats fish off the African coast now and others go up to Svalbard. They're the ones that are emptying the world's oceans."

Fishing is in Rödström's blood: his father, grandfather, great-grandfather and brothers all made their living from the industry. His great-grandfather fished cod in the 19th century and he, his father and grandfather have all done purse-seining, mackerel-fishing with drift nets, herring-fishing in the North Sea and shrimp-fishing by the coast.

"Here in Sotenäs municipality we had 150 boats in 1953, each with crews of three to seven men. Today there are six boats, each with just a couple of men. I've never been as pessimistic as I am today. No youngsters are interested in becoming fishermen; it's hard to get a permit, and it's also difficult to find reasonably priced boats. The EU is paying millions to send boats to scrap, so who can afford to buy one?"

Rödström recounts how fisheries policy works since Sweden joined the EU. No additional net tonnage is now allowed in the fleet, which is good. On the other hand skippers have been remunerated for scrapping old vessels and paid subsidies to build new boats of similar tonnage, which is bad. Numerous old Smögen-based wooden vessels have been scrapped and their tonnage transferred to new, modern vessels mostly owned by fishermen further south.

"Most of the skippers down there had herring boats originally, but when they were forced to stop fishing in the North Sea they sold them, replaced them with smaller iron boats and started fishing here," he sighs, wearing a look of resignation.

"We used to live in our own little world up here in Sotenäs. And people down in Gothenburg virtually laughed at me when I questioned the fleet modernisation. 'Wooden boats are a thing of the past,' they said. But they were perfectly adjusted to our conditions, you never needed to go more than one hour out to sea and there on the clay were always fish. Now the cod are gone. Why weren't we allowed to keep our tonnage here? We had voluntary restrictions on shrimp fishing too – we never fished more than we could sell. We now have unsold shrimps almost every week at the auction; they just end up on the rubbish dump. That never happened in the old days."

Then he adds something he obviously wants me to take notice of:
"And then there's a lot of fish that passes under the radar."
What do you mean?
"Well, the official rations are sold at the auction, but what's been

caught above the fishermen's rations gets sold directly on to fish merchants. That never used to happen in the old days either."

Without changing his friendly countenance Rödström gets up and looks through the large windows at the deserted quayside outside.

"It's so sad," he says, "so sad that there are no youngsters taking over. And I can't say anymore that I'm convinced that we will have any fisheries up here in the future. It's really sad."

There is a handful of white plastic tables in Gösta's fishmonger's, and I am sitting at one of them, consuming cup after cup of coffee. I spend my time watching Gösta and his infectious energy as he fills the glass cabinets with ice and fish, pours out pints of shrimps, writes new prices on the black board, hoses down the floor, and chats with everyone who comes in; customers and non-customers. The latter category is not insignificant at this time of year. It seems like everyone likes to drop in just to warm up and hear the latest gossip. During a brief lull, Gösta amuses himself by offering me a plate of biscuits and introducing me to his "assistant", the twelve-year-old Simon. Simon has a red apron and baseball cap, just like Gösta, and works here part-time after school. His job is to sweep the floor, fillet fish, clean the counter, and run errands. As I watch him he makes himself busy every second, seems to enjoy it too.

"Simon, oh Simon!" Gösta calls out now and then. "You Star of Hope, oh Simon! You know what, reporter, this lad is our future. You should talk to him!"

I ask Simon if he wants to be a fisherman. He nods.

"I would if the Green Party and Board of Fisheries didn't have all those rules," he says intently. "They just sit at their desks and decide to scrap all the fishing boats and I think that's wrong. Those people don't know anything, not even how to gut a cod."

Has he ever seen a boat being scrapped?

"Sure," he says. "First they take the cabin off and then break them up with a digger."

Now a man aged around forty who has been quietly reading the local paper looks up and shows sudden interest in our conversation. He is Roger, a former fisherman from the village of Tången.

"I have a son who's nearly fifteen and he's never caught a real cod. I mean, one weighing more than a kilo. When I was little we used to jig for cod and often caught fish of five or six kilos."

He gestures towards the grey sea.

"In the early 1980s all you needed was to sail half an hour out from Hållö and put out 16 nets. Each net was 60 metres long and you'd get anything from 300 to 500 kilos of fish. Ten years later you needed twice the number of nets just to get 200 kilos – if you were lucky. Now you get nothing. If you want fish you've got to go a really long way out and trawl. With nets you get nothing."

Roger folds his newspaper, stands up and puts on his cap.

"We've lived on fish up here for centuries but now the fish are gone. You can lay 150 metres of net and don't get a single plaice. No one even uses nets anymore, not even to try to catch a fish for dinner."

He nods, looks at his watch, says goodbye to everyone and heads off for his new job on land. Simon looks at the door closing after him, and tries to explain to me in dialect what kind of boat Roger used to have:

"A bottom yarn jolly!"

I'm completely lost, but Gösta comes to my rescue translating to Stockholm-Swedish:

"A small boat for net fishing," he chuckles, taking an appreciative look at his assistant and repeating his mantra while energetically cleaning the already shining counter: "Simon, oh Simon, you Star of Hope!"

The assistant himself springs into life when he suddenly sees a fishing boat heading towards the harbour. It is the Ocean LL 677, a 21-metre wooden trawler that is on its way in and Simon and I go outside to wait for it.

Simon grabs the hawser from the boat in an experienced manoeuvre and jumps on board to help unload the stacks of pale blue and turquoise trays full of shrimps.

The two fishermen Bengt Hartvigsson and Martin Boman, both wearing orange oil skins, sigh on hearing I'm a "reporter", but agree to talk after the catch has been unloaded.

"I love my job," Bengt says as we stand on deck beside the gaping hole leading down into the hold. "I've been doing it since 1977 and it was fantastic in those days. We never fished at nights or weekends. Then there were a lot of new boats built in the 1980s and 1990s and things got worse."

"But ..." he looks down warily at my notepad, adding: "It's been a long time since we caught as many fish as we do at the moment. You can write that."

"Why then, do the scientists say stocks are collapsing?" I ask?

"There is so much prestige in this now," Bengt says wearily. "But they're damaging an entire industry. I mean, we're producers: we put food on tables, pay tax and create jobs. We're the first link in a long economic chain that keeps coastal communities going."

He does not agree there is a risk of overfishing.

"No, because when some stocks get low you just move on to something else, that's what we've always done. If there are lots of nephrops, we catch nephrops; when the catch drops we go for shrimps instead. If there are lots of herring we have to catch them because they eat all the cod roe otherwise."

Martin Boman joins in, saying, "The real problem is the meddling from the Board of Fisheries. They're the regulator but they're always one step behind. Everything moves too slowly and causes a negative spiral. It's wrong that they interfere in what we do, they should leave it up to us, that would be better. The sea isn't free anymore."

Bengt doesn't fully agree:

"But there have to be rules, only it's gone too far. If I wait a couple of days before filling in a catch report I start getting letters threatening to take away my licence, fine me or even put me in prison. Sometimes they call and ask me about 20 kilos of fish!"

Their boat is still profitable, they say, because it is old – built in 1947. It may be slow but they do not have high loan repayments to meet every month.

But surely, I insist, they must have noticed the drastic decline in cod stocks?

"Absolutely. The fish don't like it as much here anymore," Bengt says. "I don't know why, but we also see them in places where they weren't before. Something in the environment has changed. It's got warmer, too."

So it has nothing to do with overfishing?

"Why have the fish disappeared close to shore then?" they counter. "There's been no trawling here!"

Bengt adds that seal and cormorant numbers are much higher now than in the 1970s. "Five kilos of fish per day multiplied by 10,000 seals is an awful lot of fish. But the seals don't have log books!" he notes dryly.

I insist again: What do they think about the research indicating that total stocks of predatory fish internationally have fallen by 70-90 per cent since industrial trawling began in the 1950s?

Both men fall silent for a moment; this is seemingly news to them.

"Sure, the North Sea is overfished but here not anymore. Most of the boats here have been sent to scrap."

OK. So you mean there are no problems at all?

"Sure there are," Bengt admits. "The government has to show they are on our side and get people to eat Swedish fish. And we need higher prices. If prices are better we won't have to fish as much as we have to do this day."

Bengt glances at 12-year-old Simon, who's been attentively listening to our conversation, and tells me that he himself is a seventh-generation fisherman and that it is the only livelihood he knows.

"I don't know what I'd do otherwise – maybe fish tourism. Perhaps I'll just have to go back to school and learn something new."

Martin jokes that they should seek refuge at the nearby animal park Nordic Ark, centre for endangered animals. "We'd be ideal specimens!"

We laugh and as I step back on to the quay the men call out, "Please give us a good write-up. The media are always giving us stick. Do something good with this, OK?"

I promise to do my best. Simon accompanies me back and as we go our separate ways, he too gives me a serious look and echoes: "Please, reporter, give us a good write-up! OK?"

Smögen's fish auction house, recently renovated with a 335,000 euro grant from the European Union, seems to be designed for future times when fish catches might be up to a hundred times larger than they are this day. It is Thursday, the day of the week when auctions take place in the afternoon, otherwise they are held every morning at eight. The building looks pretty much like a giant warehouse, with wet, grey, newly cleaned floors and fluorescent strip lights shining from the ceiling. Two small rows of light-blue plastic trays containing fish stand in the middle of the floor. In a corner white cardboard boxes full of cooked shrimps have been stacked, and it is here the auctioneer begins. I am the only outside onlooker but no one takes any special notice of me: the auction is a popular tourist attraction during the summer.

Shrimps are a problematic issue this year, since Norwegian fishermen according to the Swedes are dumping prices, going below what the Swedes can sell them for. But even if the Swedish shrimps aren't profitable on the market, they are still worth four euros a kilo in the fish auction: this is the current EU intervention price on shrimps – a system to guarantee fishermen a minimum earning on their catches. The reason for the abundance of shrimps at this moment is of course the decline in the stocks of predatory fish that normally feed on shrimp. At Gösta's I have heard the local description of the opportunistic shrimp that breed in billions if no fish controls them: "The louse of the sea". This spring over fifty tonnes of Swedish shrimp will be dyed green (to make them unsellable) and be put on the rubbish dump, and fishermen will be given 300,000 euros of taxpayers' money for their troubles in fishing them.

A young man with a laptop computer on a sort of tray hanging from his neck, gets the auction under way. Three middle-aged men are apparently the only buyers and assume their well-rehearsed auction poses – silently starting the bidding with the help of raised eyebrows and almost imperceptible nods. When a box of shrimps is sold, the price is written down and the buyer sticks a piece of paper with his name on the box. The shrimps are sold quickly and it is soon time for the trays of fish on ice lying on the floor. Some contain a single species, such as cod, others a mixture. One has some flat fish, a few small cod and a 60-centimetre spiny dogfish. This small shark goes by the scientific name *Squalaus acanthias* and looks exactly like a shark should, only in miniature: the skin grey and leathery, the eyes greenish-yellow and cat-like, the nose flat and broad, and the dorsal and tail fins characteristically shark-like. I suddenly notice its gills are moving regularly. Although it must have been out of the water for at least an hour there is no doubt it is still alive.

Seeing the small shark deepens my already strong sentiment of unease. I have just read Richard Ellis's book, *The Empty Ocean*, which describes how North American fishermen switched to fishing spiny dogfish after cod stocks collapsed in the early 1990s – without any protests from government authorities. Dogfish, which can occur in schools up to 20,000 strong, used to be seen as fit only for cat food, fishmeal and leather – but was all of a sudden being marketed as food in New England supermarkets. Of course the name "spiny dogfish" was deemed commercially unattractive, so the first step was to launch it as "Cape shark". This rebranding won little success on the domestic front though, so large quantities of dogfish were then exported to the UK, as "fish" to be sold in fish 'n' chips shops. And of course the fins could always be sold to Asian markets, where they are used to make shark fin soup.

Then the news came in 1998 that spiny dogfish stocks in the North-west Atlantic now also were overfished. That wasn't very surprising, since this species reproduces extremely slowly, with females not reaching sexual maturity until the age of 12 years and gestation lasting for 22 months, along with elephants, the longest of any vertebrate. The young – four to eight in each litter – are born live, and the female gives birth only every two years. Although this extremely slow reproductive cycle hardly came as news to the scientific community, it was not until 2000 that the US introduced any restrictions at all on dogfish catches.

(Sweden would not impose limitations until 2007, despite dogfish having been on the International Union for Conservation of Nature's Red List since 2001. The EU finally introduced a ban on spiny dogfish fishing in 2010, after a decline in stocks of almost 99 per cent.)

The spiny dogfish has a string of alter egos, including mud shark, grayfish and spurdog, and to be frank – the flesh is not very tasty. At the fishmonger's it seems to be used in the display mostly as an eye-catcher. All the more unnecessary to see how stocks in the US and Europe have followed the same curve, first rising and then suddenly plummeting – a warning sign to which policymakers have taken far too long in responding.

In front of me the dogfish is still struggling for breath. I find myself wondering how much it would cost. Twenty euros, perhaps? Even less? A probably very pathetic plan crystallises in my head before I can stop it: why couldn't I just buy the tray of fish, walk down the quay and tip the dogfish back into the sea? I note its small white spots on its sides and reflect again on how much it reminds me of a toy shark. And sure enough, this one is only a youngster; dogfish are long-lived and can reach almost forty years of age. They can grow up to 1.3 metres in length, weigh up to 15 kilos and are impressive swimmers: individuals tagged in Norway and the US have been recaptured in the Bay of Biscay and Japan respectively. Dogfish are also highly sociable. As one angler's website states, "A distinctive character of the spiny dogfish is that it often follows its unfortunate hooked comrades up to the surface. It is not unusual to see a whole shoal following, and then slowly swimming away."

But before I get any further in my thoughts, the mumbling, nodding and eyebrow-raising auction moves on – and a white piece of paper drops on to the tray with the shark.

How much it cost I never found out.

After the auction I ask one of the staff, Björn Bengtsson, to show me round the new premises. Thanks to a European Union Financial Instrument for Fisheries Guidance (FIFG) grant of 335,000 euros, the Smögen auction house has a new packaging storeroom, toilet, conference centre, space for wholesalers, and a training kitchen earmarked for "training in the fishing industry and by schools".

The FIFG has been especially generous to Smögen and the surrounding area in recent years. According to the *Investment Catalogue 2000-2006*, it awarded structural grants to ABBA Seafood in nearby Kungshamn for production plant modernisation: 50,000 for "the purchase and installation of an automated tubular packing line", more than 60,000 for "building extension and the purchase of plant to expand the product range", and 37,500 euros for "the purchase and installation of machinery". A little surprising given that ABBA Seafood is a large, privately owned company.

The FIFG also contributed 738,500 euros to build a new cold-storage facility in the area, and 2 million to scrap 13 locally based trawlers, with owners being offered between 60,000 and 260,000 euros to decommission their boats. At the start of the scrapping programme (the "structural adjustment" as it is called in EU language) the Board of Fisheries had difficulty finding willing candidates – a fact that the owner of one boat, the 32 metres long Hållö LL 149, profited from by negotiating a pay-off of 860,000 euros to decommission his 1976 vessel. A golden parachute fit for a European banking executive, you might think.

But is 860,000 euros a reasonable sum to scrap a vessel from 1976? Is it much or is it little? Scouring the pages of the *Yrkesfiskaren* fishing trade magazine, I cannot find a single second-hand boat advertised for sale at over 300,000 euros. A Danish 38-metre industrial trawler in good condition and built in 1999 is listed at 220,000.

In a dark corner of Smögen's gleaming new but empty auction house I spot the tell-tale signs of another EU project.

"This is one of those machines for scanning crabs to see if they have food in them," Björn Bengtsson explains.

The gadget resembles an outsize overhead projector and looks suspiciously unused.

"It was the idea of Bo from Kornö Island to invest in crab here and we got EU funding for this machine, but it didn't really work out. I don't know why."

Bengtsson apologises – he has to go back to work now. I follow him through the shining clean new hall. In a back room I catch sight of the spiny dogfish, its gills still pumping. Not for a second do I believe I have psychic powers though in that moment I would have bet my last penny that this species would soon be a European matter of concern. And my prediction was proved right only two years later, in 2007, when Germany proposed it be added to the CITES (Convention on International Trade in Endangered Species of Wild Fauna and Flora) list, which it was.

Germany's special interest in preserving the fish is because *Schillerlocken* (strips of smoked dogfish) are a national delicacy. The singular name comes from the fact that the shark's belly sides curl during smoking and resemble the blonde locks of the Romantic poet Friedrich Schiller. This I discover while searching the internet for dogfish, and find the German word *Überfischung* as soon as I find web pages called "Requiem für die Schillerlocken". No one can claim the fate of the spiny dogfish has come unannounced.

I leave the auction house feeling even more ill at ease. But with no premonitions about another sad fact: that I was one of the last visitors there. During the autumn the auction house closed down and was replaced by a foreign auction website. A distraught Gösta Karlsson, owner of Gösta's Fisk, told the *Aftonbladet* newspaper that one of the strongest tourist magnets in town had thereby disappeared.

"It's a disaster, it ruins everything."

Who's to blame? The authorities? The fishing industry? The politicians? Who are the bad guys, really? People outside the fishing world always ask me the same question, and I myself keep turning this question over and over inside my head. The easiest journalistic angle of them all for me would be to find the simple story about the 'Bad Guys'. The arrogant authorities. The corrupt politicians. The unscrupulous industry. Or the evil foreign interests. From a human perspective, I honestly would prefer finding someone other than the fishermen to blame, and I'm not the only journalist thinking this way. Former Board of Fisheries Director-General Per Wramner, the man with the fringe who sided with the industry and criticised professor Lars Hultkrantz's figures, told me how he perceived that the media routinely gave him the negative role:

"In dramaturgical terms it's much easier to portray men in oilskins on the boats with their broad dialects as the good guys, and me, the authority at my office desk wearing a tie and glasses, as the bad guy. Most journalists seem to see it that way."

In Smögen I feel an acute need to sympathise with the fishermen: Bo from Kornö island, who has fought and is still battling away; young Simon, "The Star of Hope", so enthusiastic and hard-working in his right element – a separate universe from my own computer game-playing son of the same age; Gösta the fishmonger, with his wonderful knack of brightening people's day; and the mild-mannered Sven Rödström. Are they the ones to blame for the subsidy deluge, the misguided investment support, the regulations that have forced fishermen to dump their catches, the rules that allowed more than 6,000 tonnes of cod to be sent to landfill in the 1990s, for the lack of rules on one side and a nit-picking bureaucracy on the other?

I cast my mind back to Bo Hansson's piles of old documents and reflect that no politician can justifiably claim that the facts were not known. Hansson showed me a 1998 performance evaluation of the Board of Fisheries, a report which drew attention to the Board's dilemma that on the one hand it should be providing services to a single, small industry, while on the other also having responsibility to regulate and control the same industry. The weighty document detailed the close links between the sector and the Board and called for firmer political governance to support the Board's officials that too often felt alone and abandoned. But little happened after the report arrived at the government offices. As Nils Gunnar Billinger who wrote the report stated later: "The fundamental problem was there was no fisheries policy. There was no policy to guide the authority so the Board was pretty confused and didn't know what to do. The staff would go back to the Ministry and ask the politicians and civil servants for help but nobody wanted to get involved. The civil servants blamed the politicians for not formulating the policy, and the politicians didn't have the time to make policy in such a minor area."

Many have pointed out to me that Sweden's disastrous fisheries policy to a large extent is due to the industry's small size and marginal economic importance. Fisheries policy has become "the small minority's hostage", as the government's Environmental Advisory Committee in *Strategy for Sustainable Fisheries* put it in 2006: "Without the corrective mechanisms that apply in areas accorded political priority by prime ministers or opposition leaders."

In countries like Iceland and Norway, where major national economic interests are at stake, sustainable fisheries are not treated as a secondary issue. Norway, which lands around 1.3 billion euros[2] of commercially caught fish per year and earns about the same amount from fish farming, has managed its fisheries far better than Sweden. Discarding of fish (dumping) is banned, so all catches must be landed. If large numbers of immature fish are caught the authorities impose a temporary fishing moratorium because catching juvenile fish wastes a precious resource. Norway's penalties for fishing illegally are also considerably higher than the EU's.

Iceland, meanwhile, has chosen an alternative route with its individual transferable quotas (ITQs). These "fishing rights" are sellable and considered by many a controversial fisheries management model. But they do have the advantage of giving fishermen an incentive to manage and take care of stocks, in the same way a farmer looks after his fields

2   NOK 10 billion.

(more of this in chapter 10).

So who profits from overfishing Sweden's seas? Surely *someone* must, otherwise the relentless opposition of commercial fishermen to most regulations and scientific recommendations would defy logic – as would their failure to agree on voluntary management that would benefit all.

To understand the driving forces behind overfishing I decide to find out who the dominant players are in the Swedish fishing industry, and how much money is actually at stake for those who fish the most.

I will be very surprised.

THE SUBSIDIES

Lift up your eyes to the heavens,
look at the earth beneath;
the heavens will vanish like smoke,
the earth will wear out like a garment
and its inhabitants die like flies.
But my salvation will last forever,
My righteousness will never fail.
(Isaiah 51:6)

Whoever saw the documentary "The Last Cod" *(Den sista torsken)* by Swedish filmmaker Peter Löfgren on Swedish television in 2001, might remember the scene when fisherman Tore Ahlström, aboard the newly built, 45-metre trawler *Tor-ön* (one of the boats Sven Rödström was referring to when he spoke of boats worth tens of millions of euros), talked about his belief in the Almighty. His family, the 17 people listed as the owners of four of Sweden's largest fishing boats, is part of the group that has consistently been the first to invest, innovate, build new boats, fish new species, seek new waters. The family made its fortune during the Baltic cod boom of the 1980s and ploughed back most of its profits into bigger boats. In addition to this the family received one million euros[3] from the munificent EU Financial Instrument for Fisheries Guidance (FIFG), as well as receiving Swedish state subsidies to build the industrial trawlers *Ganthi* and *Ginneton* in 1996, so the family could target sprat for fishmeal in the Baltic when cod stocks were depleted.

---

3   All the amounts in euros are based on the simplified exchange rate of 1 euro = 10 SEK. The rate has varied over the years somewhere between 1 euro = 8.8 SEK and 1 euro= 11 SEK.

"Thou shalt invest thy pound," fisherman Tore Ahlström said in the TV programme. "Wise investment is a gift from God. It's against God's wish to lay up thy pound in a napkin. We are to leave a legacy of prosperity, as our ancestors did."

The fact that the Ahlström family belongs to the evangelical church in Fiskebäck is well known, and in the programme Tore is happy to talk about his belief in God and his positive thinking as keys to success. But towards the end of the documentary, his optimistic attitude is unexpectedly punctured: pressed by Peter Löfgren on the decline of cod and the risks associated with industrial fishing Ahlström suddenly claims that overfishing had been predicted in the Bible. All of it – "the thinning of the ozone layer, environmental destruction, overfishing, everything" – was referred to in the Bible by Isaiah and in the Epistle to the Hebrews. Heaven and earth "will wear out like a garment".

"Don't ask me how," Tore muses. "But I can see that it is being worn out like a garment. For a believer, this is not a surprise. Everything is *supposed* to end; everything is *supposed* to wear out."

This TV clip has had ramifications in fishery circles in Sweden: everyone – anglers, fishery biologists and fishermen – inevitably react with horrified shudders when talk turns to the Fiskebäck fleet. But I, unfamiliar with the real workings of the fishing industry, am still unable to judge whether the Ahlströms, with their four big boats and their frighteningly literal Biblical interpretation, really are the major actors in Swedish fishing or not.

To find out, I turn to the Swedish Board of Fisheries. But the issue turns out to be more sensitive than I thought. The official I first quiz about the major fishing companies is initially very hesitant. The data are probably confidential, he ventures, "Since it concerns individual people".

But following a few more enquiries, where I challenge the confidentiality status given to "individuals" freely fishing a common resource, who receive allocations from a politically based quota and are in addition supervised by public authorities paid for by taxpayers' money – the information is emailed. It's the list of the top twenty cod fishing vessels and similar lists for the pelagic fleet – i.e. sprat, herring and fishmeal vessels.

Initially I am disappointed at how little information I can glean from the cod fishermen's list; my interest in overfishing had first been aroused by the plight of the cod. But no giant boats, no boats with any apparent links, no vessels with sensational catches catch my eye. The ten biggest boats all seemed to be making roughly the same money; all with revenues of between 400,000 – 500,000 euros and owner profits of around 15,000 – 30,000 a year.

But then I realise why: cod fishing is no longer the big money-spinner it once was. Of the value of just under one hundred million euros for all landed fish in 2005, cod represented only just over 16 million euros compared to 1990 when cod brought in 44 million. In other words, the big players have deserted the declining cod and invested in the more profitable "industrial" species: herring, sprat and other fishmeal catch, to a combined value of approximately 37 million euros.

My initial disappointment is soon superseded by an interest in what are known as pelagic ships – the herring and sprat fishing trawlers. Immediately I detect an apparent imbalance: of the 195 vessels fishing herring and sprat, the ten largest were taking as much as 30 per cent of the catch. And of the top six, I recognise four names: the Ahlström-owned *Tor-ön*, *Torland*, *Ginneton* and *Ganthi*. The same boats also appear in the top ten on the fishmeal list, where the ten largest vessels account for over 40 per cent of the catch.

My next step is to find out how much money is being made by these big boats and I call the Companies Registration Office. Here the figures are delivered without delay, due to the Swedish Freedom of Information Act. I am told that two companies Ganthi Ltd. and Ginneton Ltd., named after the boats, were each turning over more than 2 million euros, although the owners' taxable incomes were rather modest: between 20,000 and 30,000 euros.

But what about the two newest and biggest ships in Sweden then, the *Tor-ön* and *Torland*? Here I run into some trouble, since the Torönland company that owns both boats had not submitted any annual reports to the Swedish Patent and Registration Office. This was completely legal since only trading companies with a turnover exceeding 3.8 million euros are required to do so. Lacking revenue data, I decide to check owner incomes. Torönland turns out to have nine owners listed, all with the same surname: Ahlström.

I call the Tax Agency to find out the incomes of these nine people, expecting to be surprised. I am frankly hoping for huge, staggering sums, incomes to compare with the tens of thousands of tonnes of silvery herring and sprat vacuumed up by their giant boats each year. But the surprise is of another kind. My contact at the Tax Agency runs the figures again and again but keeps coming up with the same result: zero income. None of Torönland's nine owners reported any income at all, and not just for one year but for six consecutive years: 2001-2006. In other words, these nine members of Sweden's most prominent fishing family were seemingly broke. For the 2005 financial year, the nine Torönland executives had not only a combined income of zero kronor or euros but more than 800,000 euros in losses from active enterprise. And for the 2006 financial year their combined business deficit was a dizzying 1,215,800 euros.

I recall some dialogue from Peter Löfgren's film, "The Last Cod", portraying the religious Ahlström fishermen.

"So how's the fishing going?" asks Löfgren. Tore Ahlström leans back in his chair on the bridge of the *Tor-ön*, just having explained that the two boats take about fifty hours to fill, and that a single day of trawling can net them 300,000 kilos of fish – enough to fill ten transport trucks.

"We're completely satisfied," Ahlström replied. "It's fantastic the way it's been going this spring – well, I tell you: you couldn't wish for more!"

I continue my research, looking at other fishermen's incomes, but the owners of the other most successful pelagic boats are not reporting such jaw-dropping negative income statistics. Remarkably enough though, only three of the people on the list reported incomes that could be described as substantial. These are the Johansson father and sons, owners of the big modern *Astrid* and *Astrid-Marie*, vessels built with a grant of over 900,000 euros from the Fund for Fisheries Development and public money in 1996. Their reported annual taxable incomes were 40,000 to 140,000 euros for the years I check.

From my lists of Sweden's busiest fishing boats, it turns out I find only one person with an income of over 100,000 euros a year, but many more not reporting any taxable income at all, and for the rest, people with normal or quite modest incomes. I cannot find any numbers whatsoever to indicate that this is a gold-digging, money-spinning business, worth fighting tooth and nail over. Going by reported incomes, most of these people should rather be racking their brains over how to get into more profitable trades!

So why are the Ahlströms still fishing you might wonder? An intriguing detail in the story is that the two boats that apparently were not making a single Swedish krona and instead leaking money big time, were among the last fishing vessels to be built in Sweden and, in contrast to the almost equally large *Astrid* and *Astrid-Marie*, they were built without EU subsidies. That was because the Board of Fisheries didn't give its consent since it judged there was no more room for large boats in the pelagic segment of the Swedish fleet. But several sources say that the Ahlströms forged ahead anyway, since their main competitors, the Johansson family, had been given EU grants to build their boats.

"It's like when everybody else is pregnant, you develop a phantom pregnancy," joked a senior official at the Board.

So – the Ahlströms dug deep into their own coffers and then went to the bank in Gothenburg. The bank lent them 7.5 million euros to build the *Tor-ön*, and 8.2 million to build the *Torland*. On top of this, the family was lent 800,000 euros[4] by Danish FF Skagen, one of the world's leading producers of fishmeal and fish oil.

And as a consequence – the family is now dashing around in Swedish waters sucking up herring and sprat solely to meet the interest payments on their loans, and without making any money for themselves at all. Perhaps it was this economic dilemma that made the family shut off their VMS transmitter on at least one occasion, when they were caught fishing illegally. The skippers of the *Tor-Ön, Torland, Ganthi* and *Ginneton* were convicted a few years later of having stolen 239,000 kilos of herring and sprat from a banned area in the North Atlantic. The four skippers denied the charges, but the judge attached no credence to their claim that the VMS transmitters of all four boats had stopped working and then, by coincidence, had started up again simultaneously a day later. Neither did he believe that their catch was Baltic herring as the men maintained.

---

4   1 euro = 1.34 DKR .

Fisheries Board expert Bengt Sjöstrand had seen the catch and was certain: the four boats had been fishing autumn- and winter-spawning North Sea herring that was just about to spawn when illegally caught. Had the herring been from the Baltic, where it spawns in the spring, it would have been "a scientific sensation" Sjöstrand told the court.

The value of the catch – 106,900 euros – was confiscated by the Gothenburg district court, which also slapped the skippers with maximum fines totalling 72,200.

The court of appeal denied the skippers an appeal hearing and the fines have, surprisingly, been paid: at any rate, there is no record of outstanding debts to the authorities.

But loans to the bank still need repaying. The Ahlströms' desperation was tangible also in another district court case: the family sued the Fisheries Board for not issuing the *Tor-Ön* and *Torland* catch quotas in the Norwegian Sea, where Sweden has fishing rights. The plaintiff's outraged tone is the same you always hear in fisheries debates: boats were not being given proper "justice": their quotas were "a slap in the face" compared with what others got: they were "rigorously pursued" by their own trade union SFR, whose boss, Reine J. Johansson, "runs the Board of Fisheries".

But the family lost this case as well, and, as a consequence, they felt forced to seek fishing waters further abroad. Oblivious to public concern, the family stationed one of its biggest boats, the 50-metre trawler *Ganthi*, in controversial fishing waters off Moroccan-occupied territory of Western Sahara. This made the Ahlströms complicit in a fishing operation strongly opposed by the Swedish government and the UN, and categorically condemned by human rights groups as a breach of international law and theft of natural resources from an occupied people.

"Swedish politicians have betrayed its own fishing industry," countered skipper Ove Ahlström in the press (*Göteborgsposten*, December 5, 2005). "It wasn't our choice to go to Western Sahara – Sweden has a Good Neighbour Agreement with Norway that is not adhered to. The Swedish government's fishery policy is what makes Swedish fishermen go abroad."

But despite what surely must have been a lucrative gambit, the *Ganthi* reported diminishing returns after the move to Africa – from 2.7 million in 2001 to 760,000 in 2005 – while its sister ship the *Ginneton*, fishing home waters as usual, reported revenue between 1.7 and 2.7 million.

In 2006 *Ganthi* was sold in a non-EU country. The Ahlströms brought in a new vessel of equivalent tonnage, called the *Ganthi 8*.

"It represents an increase in fishing capacity," noted an official at the Board of Fisheries. "Even if the tonnage is the same, a new boat is always more modern and more effective. It's more comfortable, has improved refrigerated space, better pumps and better navigation equipment. All in all, this makes it easier to harvest more fish."

"You must understand that fishing isn't about economics, it's about culture," a fisheries manager explains to writer Charles Clover in his book *The End of the Line – How Overfishing is Changing the World and What We Eat*. This is when he wonders why Canadian coastal fishermen keep on claiming unemployment benefits year after year, still waiting after 14 years for the cod to return. Visiting the small coastal settlements, Clover noted that fishing is not like other industries. It is hunting, it's a lifestyle, it's being in touch with the elements. It is skill and luck, and the dream of a giant catch, it is an independent life with freedom to sail or take a break –there are a thousand reasons to want to stay in the business if you get the chance, even through a few lean years.

And many get the chance – far too many. Taxpayer money has helped numerous fishermen across the world to maintain a lifestyle and culture that fish stocks can no longer sustain. The Canadian authorities were so hopeful that the cod would return that they undertook to support 44,000 people for more than two years. The so-called Atlantic Fisheries Adjustment Package cost, according to the Canadian audit office, C$4 billion between 1992 and 1995. The result of this "adjustment" was, the audit office said, that a couple of years after the cod had collapsed, a huge fishing fleet was equipped and ready to resume fishing, with newly renovated boats and computer-trained, rested manpower – and with capacity at 160 per cent of its 1992 level. The problem was that the cod had not returned. And it still has not in 2011...

In Sweden too, as previously mentioned, public funding has been lavished on the fishing industry; in fact, in inverse ratio to the amount of fish in the sea. Sweden's entry to the European Union in 1995 gave the fishing industry "structural funds" amounting to 150 million euros up until 2006: a third from Swedish taxpayers and the rest from the EU.

According to the Board of Fisheries report *Fuelling Fishing Fleet Inefficiency*, the money has increased capacity in the so-called pelagic segment (herring and fishmeal trawling) by 50 per cent, while catches have dropped by 20 per cent.

And what other public money is propping up the fishing industry? On top of the annual 13.5 million euros or so from EU structural funds, several other public outlays are channelled into the industry, though it is arguable whether they all can be labelled "support".

One of these is the annual cost of running the Board of Fisheries (24.4m in 2006). Of course you can debate whether or not the authorities supervising the industry is a "support", but it is still worth discussing since the Board's administration and research departments are explicitly required to work for the fishing industry and fishing management, which undeniably limits their mandate. The head of the Board's inspectorate, Johan Löwenadler Davidsson, once stated this as unequivocally as possible in an interview: "We exist for the fishing industry, nothing else. That must colour everything we do."

The Board of Fisheries' three laboratories – the Marine, Coastal and Freshwater laboratories – have all worked for decades within that definition. In recent years, the labs have produced valuable fundamental research about the cod and other commercial fish, but throughout the 20th century the Board's laboratories have rather acted as a brake on research by other scientists into the role of the major fish species in marine ecosystems. Sture Hansson, a professor of ecology at Stockholm University, ran into a brick wall when applying for funding to research the cod in the context of the Baltic Sea's marine ecosystem. He was repeatedly denied any money and was told that research into commercial species was the exclusive domain of the Board of Fisheries' laboratories; no other institutions ought to get into that.

Those that want to make the argument that the Board's budget is in fact largely dedicated to research and conservation of marine resources should therefore ask themselves the following: what if the three laboratories had been attached to universities instead? Or to the

Environmental Protection Agency (EPA)? The coastal fisheries' lab in Öregrund actually did fall under the EPA once, but was transferred to the Board of Fisheries in the early 1990s, at a time when environmental conservation needed it the most. Would the EPA really have neglected to test-trawl along the west coast for 20 years when fishermen and the general public were warning that cod were disappearing? Would the EPA have needed approval from some "resource section" at the Board of Fisheries in Gothenburg to carry out such research? If the laboratories had belonged to universities, would Sweden's marine biologists who were not employed by the Board of Fisheries have allowed their research to be limited to plankton, bladderwrack, rock cook and the sex life of sea horses, letting the fishing sector's own administration have exclusive rights to research the most abundant species in the marine environment, the ones which impact the ecosystems and water quality most of all, namely cod, herring and all the other so-called "commercial" fish?

Let's face it: the Board of *Fisheries* exists to serve the *fisheries* sector. It serves under the Ministry of Agriculture, not under the Ministry of the Environment. Fishing used to be seen – and mostly still is – as food production, like agriculture. And just as nature conservation is not the Ministry of Agriculture's main task, neither is protecting fish and marine environments the Board of Fisheries' principal job. In conclusion: the 24.4 million euros a year that the Swedish Board of Fisheries cost can to a large extent be attributed to the fisheries sector.

Another, indirect, support for the fishing industry is the cost of unemployment benefits for fishermen. In 2006, this surpassed 2.8 million euros for the 862 recipients, meaning that the overwhelming majority of the 1,100 people registered as professional fishermen were on benefits for an average of 55 days that year.

This is less a regular payout to the jobless than an industry subsidy since registered fishermen are not obliged to seek other work or close down their companies to get benefits, in contrast to other industries. This was pointed out in a Ministry of Finance-backed report written by professor of economics Lars Hultkrantz and associates. Fishermen, even those running their own businesses, can claim unemployment allowances for days off because of bad weather, ice, breakdowns, emergency boat repairs and even because quotas have been reached. These days around 50 per cent of all benefit payments to idle fishermen come in connection with imposed bans on fishing. This means fishermen that under normal market economic rules would be forced to look for other employment when fish stocks are under pressure, are instead allowed to stay in their trade and claim benefits while at the same time the authorities close down fishing to protect stocks.

Most of the rest of the entitlements are for *force majeure* incidents such as storms, ice and breakdowns. Very little goes to what is normally understood as unemployment. That the fishermen want to keep the system this way is not strange considering the questionable future that lies in wait for the fisherman who gives up, folds his company, and joins the ranks of the unemployed. Instead of taking money and doing other chores at home when fishing shuts down by order or because of a brutal winter – he would have the questionable pleasure of having to go looking for a new job. It is no wonder that many choose to stay on paid leave, hectoring decision-makers to let them fish again at the first sign of fish stock recovery. As Jack Rice, of the Canadian Department of Fisheries and Ocean, puts it: "We can keep on until hell freezes over, producing simulations of how fish stocks would double if we held off fishing for another five years, but it won't help. People always want to begin fishing *now*."

Another form of direct financial support to fisheries comes from the EPA as compensation for damage caused to fishing equipment by seals. In 2006, Swedish fishermen received 1.7 million euros for this and an additional 530,000 euros for preventive measures such as acoustic "scareseals" and pontoons that keep seals at bay, as well as stronger yarn for nets. This type of money for actions to prevent seal damage has been paid regularly since the late 1990s, though seal damage costs have not diminished. On the contrary, in 2006 the Swedish Fishermen's Federation (SFR) demanded an increase to 5.6 million a year, claiming that 1.7 million euros does not come anywhere near meeting the need.

Other EPA project money connected to fisheries includes 300,000 euros for research into seals and fishery and 100,000 for research carried out by the Board of Fisheries into the decline of pike and perch in the Baltic (due, according to current research, to the sea's "flipped" ecosystem: overfishing of cod has altered the balance between species).

The Swedish EPA estimates that another 2–2.5 million goes to measures that indirectly benefit fishing: maintenance of water courses, research into ecosystem-based management of lakes and water courses, restoration of lakes, and the creation of migration paths for salmon and seagoing trout. But a certain number of these projects are driven by nature conservancy and also benefit angling and so cannot be defined as costs for fisheries. Costs *directly* related to professional fishing in the EPA budget amount to 2.63 million euros.

The coastguard devotes much time to fishery inspection. Exactly how much is hard to say: the coastguard's own, cautious estimate is that 9.9m of its 63.8 million budget – a sixth of its resources – can be directly related to fisheries.

The coastguard says it devotes 24 per cent of its time to fishery inspection and that the frequency of inspections both at sea and at dockside has increased dramatically in recent years, but the argument that we need an effective coastguard even without fisheries makes it reasonable to accept a low figure like 9.9 million. It is obviously wrong to call this money support for the fishing industry, but few fishermen would be prepared to forego having a control mechanism, which acts as a guarantee against anarchy at sea. Without the coastguard's random inspections, cheating would presumably be so widespread that fishermen would soon be without any income at all.

The effect of inspections on respect for the law has been demonstrated in a report "Swedish Fisheries Inspection", which noted that coastguard inspections actually improve fisheries catches. One assessment claims that average catches from inspected boats are between 10 and 50 per cent greater than on other boats.

"The question is whether it wouldn't be productive for some ships to have an inspector on board – they'd be getting more fish," as a coastguard representative put it.

Another subsidy for the industry is the EU "intervention" system, the lowest-price guarantee that lets fishermen in times of unusual abundance of certain species avoid having to slash prices. Fish are thus rather thrown away, or ground into fishmeal, instead of fishermen stopping fishing when there is an excess of a particular species on the market. In the first two years of Sweden's EU membership (1995 – 96) this guarantee allowed fishermen to toss 5,000 tonnes of top-quality cod on the rubbish heap because of low demand. (At the time, the processing industry did not have the capacity to freeze that amount.) The handout from the EU to Sweden to destroy this fish was 2.4 million euros.

Since that time, intervention price payments have varied, but the 1995–2005 average has been close to 400,000 euros a year.

For the sake of clarity, it should be noted that this money is separate from the 13.5 million euros the EU annually doles out in structure payments to the Swedish fishing industry.

Sweden's county administrative boards run several operations connected with the fishing industry and angling. The Hultkrantz report from 1997 checked the annual reports from all the county administrative boards and found that, after deductions for costs of management of angling and fishing for private consumption, 3.13 million was spent on actions relating to the fishing industry. The report also calculated the cost for the Ministry of Agriculture's own fisheries unit, and pegged it at 250,000 euros for 1996. According to a more recent count from the ministry, the figure was 590,000 euros in 2006.

A large sum that can be viewed as a fishing subsidy but not taken up in the Hultkrantz 1997 Finance Ministry report relates to the compensatory release of fish. Hydropower companies are obliged to release salmon, sea trout, eel and other species to compensate for the closure of natural migration paths in the rivers. The Board of Fisheries has no cost estimates for this but according to Sweden's major power companies (Vattenfall, E-on and Fortum) the figure is in the region of 10 million euros a year. This is for the release of salmon and salmon trout smolt and catch-ready fish, the transport of eels, the construction of salmon ladders, fishing fees and the maintenance of water courses.

Naturally, restocking is not an obvious fishing subsidy as it can be seen as beneficial not only to fishermen, but also to consumers. And the hydropower companies are of course responsible for the inability of the fish to migrate and spawn in their natural way. But it is still worth noting that consumers, through their electricity bills, are paying for the release of fish that professional fishermen can catch and then sell to them.

It is difficult to think of any other industry that so confidently claims compensation for income downtime. Every imaginable intervention in the marine environment – such as the construction of the Öresund Bridge linking Sweden and Denmark, the laying of the North Stream cable to Poland, or the planning of wind farms in shallow waters – brings a new demand for compensation. And usually they are successful. Almost every obstacle to fishing, including seals, bad weather, and even the industry's own overfishing, currently merits some kind of reimbursement.

Another hidden subsidy that the Hultkrantz Finance Ministry report didn't bring up was the fishermen's tax breaks for fuel. Swedish fishermen use about 60 million litres of marine diesel every year, according to the Swedish statistics agency Statistics Sweden.

In 2006, consumers were paying 0.95 euros a litre for this fuel, including environment, carbon and value added taxes (VAT). But fishermen with vessel permits are exempt from all such taxes, and through their companies can deduct VAT, cutting their price to just 0.4 euros per litre. Therefore exemptions from fuel taxes, amounting to approximately 33 million euros, must also be added to the list of subsidies to fishing – about 21.6m if we don't count reclaimed VAT, which any VAT-registered business has a right to claim. In the EU's Green Paper on Fishing Reform, this fuel tax exemption is counted as a "taxation subsidy".

In fact it is not unlikely that this tax-exempt fuel has played a key role in facilitating overfishing around the world. In 2007, Swedish research scientists Ulf Sonesson and Friedrike Ziegler at the Swedish Institute for Food and Biotechnology in Gothenburg showed that it takes an average of eight litres of diesel to produce one kilo of edible nephrops (Norway lobster) – clearly an unprofitable exercise if the fuel wasn't so cheap. From a global warming perspective, trawling for nephrops holds the record for emissions of greenhouse gases per unit of produced foodstuffs: 32 tonnes of carbon dioxide for every tonne of edible nephrops – more than double the figure for second-placed beef. Nephrops trawling is thus one of the most energy-guzzling food producing methods on the planet, made possible only through a magnanimous tax policy that has never previously been questioned.

While on the subject of hidden subsidies, mention must be made of the perennial argument over a special tax break for professional fishermen. For more than ten years, the Swedish Fishermen's Federation (SFR) has stubbornly lobbied for special tax relief for professional fishermen patterned on a model used in Denmark and Norway. The original proposal was for a standard deduction of 4,500 euros on income tax to compensate professional fishermen for nights spent away from home and other incidental costs. Many Swedish parliamentarians of various parties have backed this, among them the current Minister of Agriculture Eskil Erlandsson, who submitted formal proposals on the matter while still being a backbench Member of Parliament.

Ultimately, a Swedish government proposal was sent in 2006 to the European Commission, but was rejected on the grounds that it was incompatible with the common market.

Thanks to that EU principle, the deduction was not approved. But it is of interest because, had it succeeded, it would have been a kind of hidden subsidy, exactly like the industry's fuel discount, a type of policy the Swedish government claims to oppose – since it thinks all subsidies should be open and transparent! It will be equally interesting in the future when the discussion on emission rights for transport is widened to include boats. If fishing fleets carried their own environmental costs for emissions, the standard environmental taxes would be nowhere near enough. According to the report *Internalisering av sjöfartens externa kostnader* ("Internalisation of the External Costs of Shipping") by Per Kågesson of Nature Associates in 2000, the cost to society of the Swedish fishing fleet's carbon dioxide and nitric oxide emissions is at least 27.5 million euros.

If the industry paid for that, then what would happen to the price of nephrops?

So, finally, what *was* the total cost to Swedish taxpayers of the fishing industry in 2006?[5] Note that some of the following items can be disputed (arguments mentioned above).

EU structural fund: €13,500,000
Unemployment benefits: €2,100,000
Swedish EPA: €2,650,000
Coastguard: €9,900,000
EU intervention price guarantee: €400,000
Board of Fisheries: €24,400,000
County administrative boards: €3,130,000 (1997 figure)
Agricultural ministry fisheries unit: €590,000
Hydropower companies' release of salmon, salmon trout and eel: €10,000,000
Fuel tax discount: €33,000,000

Total public costs and subsidies: €99.65 million.

By this calculation, Swedish society's total costs and subsidies are perilously close to the ex-vessel value of Swedish-caught fish: approximately 100 million euros, and the amount grossly exceeds the value added for Swedish fish according to official government accounts: 58.5 million in 2006.

Of course, someone reading this may want to adjust the figures, perhaps by removing the entry for the power companies' release of salmon, halving the cost for the Board of Fisheries, and excising the entire entry for tax-discounted fuel. Figures can always be tweaked.

Green fundamentalists, on the other hand, would probably add the environmental costs of greenhouse gas emissions and hike the cost for the coastguard's services to a quarter of its budget (the coastguard admits it devotes almost a quarter of its time to fishery inspection), and slap on a few more million to the EPA budget for items associated with fishing. In other words, it is easy to arrive at completely different figures according to how the issue is to be presented.

---

5   All the figures are translated from SEK according to the approximate exchange rate of 1 euro = 10 SEK.

I have submitted, as openly as possibly, what I consider a reasonable method for adding up public costs incurred by the fishing industry and would like to add that it is an industry which most of us taxpayers would like to see survive and will happily contribute to. As long as it lives up to its part of the deal: delivering good, nutritious fish, today and in the future.

"Fishing isn't about economics, it's about culture," fishery administrator Jack Rice claimed in Charles Clover's book *The End of the Line – How Overfishing is Changing the World and What we Eat*.

The Swedish trade journal *Yrkesfiskaren* had some other words of wisdom, albeit with the completely opposite message: "Professional fishing is not an end in itself. We have society's mandate to deliver fish to the consumer," in the words of Charles Olsson, shrimp fisherman of Grebbestad.

Olsson is quite right of course. Society has given its mandate to fishermen to catch fish and invests almost 100 million euros annually to see that carried out. The tragedy is only that it still doesn't work very well. Perhaps exactly *because* of those millions that focus too much on the "culture" of fishing and too little on the right of consumers to have fish that are not threatened by overfishing on their tables.

Another item that could be pinned to the list above is the negative figure for how much we citizens actually are *losing* annually because of the entrenched abuse of fish stocks in our seas. According to an estimate by the Swedish Board of Fisheries (*Havet*, 2007) we might be talking about between 100 and 200 million euros a year, in Sweden alone. Globally, according to a World Bank and UN estimate, the world is losing 50 billion US dollars (*The Sunken Billions*, 2008) a year because of mismanagement of fish stocks. Clearly it is an economic, social and environmental bankruptcy of colossal dimensions.

So – the concept of "culture" has been allowed to rule, and normal economic laws have obviously been ignored. Destructive politics have been allowed to continue because their results have been hidden from the public gaze and no normal person could even suspect that such utterly senseless devastation of nature and culture could be allowed in a rich and supposedly eco-friendly European country such as Sweden. If this had happened on land, there would have been a public outcry a long time ago. Never would we permit tax-subsidised lumberjacks to fell immature trees in the public domain; never would we let tracts of forests be destroyed because they were "unprofitable" in the short term; never would we allow all other species to be bulldozed and killed to make it easier to reach only one particular sort of tree, and never would we stand quietly by to watch squirrels, birds and large mammals be slaughtered so that a certain profession, using tax-subsidised forestry machines, could retain its "lifestyle".

Nobody would give a moment's notice to the concept of "culture" in this context.

It is high time, really high time, to look at the alternatives. The questions are: how can we citizens reclaim our property? How do we get value for our money? How can we save the fish that belong to us, to our children, and to our grandchildren?

THE FISHERIES AGREEMENTS
WITH THE THIRD WORLD

Cape Verde, March 2006

"I'm a good fisherman," he says, and suddenly he straightens his back, sitting there at the table. For some time now, my companions and I have felt slight unease in this man's presence; without asking for permission he just sat down at our restaurant table – a slim, dark-skinned man with a peculiar, sad expression on his face.

We are in the town of Mindelo on São Vicente, one of the ten islands that constitute the tiny African country of Cape Verde. The few people that have heard of Cape Verde generally know it's the birthplace of the legendary morna singer Cesaria Evora. In Sweden people also may know it's the birthplace of the father of former Celtic/Barcelona football star Henrik Larsson.

Sweden's fisheries policy is now the EU's fisheries policy. And the EU in turn is the world's third largest fishing power, whose actions have far-reaching consequences all around the planet. And as I sit on a small spot on the globe, far away from Sweden and trying to enjoy my holiday, the European Common Fisheries Policy all of a sudden stares me in the face.

His name is Carlo. He looks genuinely sad. He is tall and thin, and has the light brown skin colour that is so typical of these islands where most natives descend from generations of mixed relations between European colonists and former African slaves. He has a thin black beard and beautiful dark eyes encircled by long, upturned eyelashes. Most noticeable about his eyes though, is that other thing: the lack of joy in them. Carlo smiles and tries to catch our attention with snatches of phrases in Danish and Swedish, but still he only exudes sorrow. His interest in us is soon explained by the fact that he worked as a temporary crew member for several years on board the *M/S Gunilla* – a three-masted barque that is Sweden's largest sailing ship. The vessel, used for training, regularly visits Mindelo harbour. Carlo tells us he really enjoyed his time on the ship and we wonder if his sadness is due to nostalgia for those times – an impression that is reinforced when he starts singing Danish sea shanties and tells us the crew used to call him Karlsson.

But when we ask what he does for a living nowadays, another story emerges. He is some sort of a guide, he vaguely explains, but soon adds: *really* he is a fisherman.

"I sold my boat last year, though. I saw no more future in it. It all began three years ago when the fish between the islands started disappearing. Before, normally you could go out at dawn and after just a few hours the boat was always full. But then it started taking longer and longer to fill the boat, and in the end you had to stay out the whole day. I just couldn't earn enough money from it anymore."

Carlo had a 5.7-metre motorised boat, he tells us. Although usually he tried to sail between the windy islands because diesel was so expensive, every trip cost him about €15. Fuel cost one euro a litre, bait fish cost, ice cost, and then in addition he had loan repayments on the boat. Just to cover these costs he needed to catch at least ten kilos of tuna every trip, as the price in the local market is around 1.50 euros per kilo. Fewer catches than that meant losing money on every trip.

"My father, grandfather, me and my brother, all of us have lived from fishing. Between these islands the tuna used to migrate, but they don't anymore, ever since the Spanish longliners from the EU started fishing here. Foreign fishing agreements are bad business. I don't know who benefits from them; but we, the local fishermen certainly don't."

The Cape Verde islands are situated in the Atlantic, 450 kilometres west of Senegal and an hour or so south by air from the Canary Islands. A former Portuguese colony, the country has never really found its feet since becoming independent in 1975. It remains poor, with no natural resources apart from stone, salt – and fish.

Despite their name, the islands are anything but green. We have travelled for hours on dusty, bumpy mountain roads in a landscape that most of all resembles the surface of the planet Mars: mile upon mile of red stone dust and barren red hills. Alongside the long cobblestone roads you see the occasional lone woman or child on foot, usually carrying bundles of twigs or tubs of water on their heads. Beyond them, there is nothing apart from the odd goat and a few eucalyptus trees defying the heat.

According to the World Bank, this is a country where 30 per cent of the population is below the poverty line, 20 per cent is illiterate and 26 per cent is unemployed. People are emigrating from the islands in a steady flow, especially men, who spend years abroad working and sending money home to support their families. The outside world mainly cares about Cape Verde for three reasons: its strategic location (which makes it especially suitable for NATO exercises), its mile after mile of unspoilt beaches (which attract foreign developers who build all-inclusive hotels at breakneck speed) and lastly – its fish. The islands' exclusive economic zone (EEZ) of 200 nautical miles full of tuna and swordfish is a veritable gold mine for those with large enough fishing boats to fish far out at sea.

Unfortunately, poor Cape Verde does not have any of these. But others have – for instance the EU. Decades of subsidies have left the EU with a fleet overcapacity of between 40 and 60 per cent (a number still true in 2011), which has, as we already know, led to overfishing of its own seas. The EU depends on supplementary fishing grounds for its fleets to operate. This was already true when Spain and Portugal joined the union in 1986, and they needed their long-distance fleets to go on fishing outside EU waters. That is basically why the EU over the years has signed fisheries agreements with more than 20 countries, mostly from the developing world. Cape Verde is one of them.

Since its first fishing protocol with the EU in 1990, Cape Verde has allowed European fishermen to exploit its only abundant natural food resource for a relative pittance. The country, which imports 90 per cent of its food, sells the fishing rights of 5,000 tonnes of tuna a year for €385,000. The European Parliament acknowledged the benefits of the deal in 2004, noting in a report: "The agreement is clearly advantageous for the Community in that the value of catches exceeds the cost of the protocol ... The average commercial value of tuna is more than €1,000 per tonne."

Thus, the EU pays 65 euros per tonne for a resource conservatively valued at 1,000 euros. An "advantageous" deal indeed – and even lucrative for the 84 Spanish, Portuguese and French ships that fish in Cape Verdean waters thanks to cheap licences. The fishermen themselves pay just 35 euros per tonne of caught tuna.

Of the various tuna species in the world (including yellowfin, skipjack, big eye and albacore), the most valuable is the threatened bluefin. Bluefin tuna are not usually caught off Cape Verde but any fisherman lucky enough to catch one can make a fortune on a single fish. In January 2011 a single 342-kilo bluefin tuna fetched a staggering US$ 396,000 (US$ 1,158 per kilo) at Tokyo's Tsukiji fish market.

And yet, the most problematic part of the EU agreement for Cape Verde is not the economic shortfall in terms of national income, but the negative impact the foreign fleet has on the islands' small-scale fishing industry (where boats are typically between four and eight metres and less than half of them have motors). Coastal fishing employs 11,000 people out of a small population of 490,000 and is undoubtedly a key economic activity that provides livelihoods for many thousands of families and accounts for an important part of the islands' fresh food supply. And, surely – fishing could do much more for the country if Cape Verde kept hold of its rich maritime resource for itself.

Unfortunately, the trend is the complete opposite. And this is where our new friend, 29-year-old Carlo Delgado Pinto, who unexpectedly sat down at our table, comes into the picture.

Seeing that he has now caught our attention by telling us about his experiences as a fisherman, Carlo suddenly has a new posture. His eyes have lit up; his previous ingratiating manner has completely evaporated. He shakes his head seriously and looks concerned when he gives us yet another reason for him quitting fishing:

"I started taking bigger and bigger risks, both financially and with my own life. The boats here are built to operate between the islands, you know. If you go past Santo Antão it is completely different sea, there are strong currents, waves; it is very dangerous to go out too far."

Most Cape Verdeans cannot swim, he says, and fishermen are no exception. When the fish disappear from the coastline, many poor fishermen are forced to risk their lives further out to sea, just the way he himself did.

"I didn't want to stop fishing and thought I'd never have to, but in the end I just couldn't go on."

He smiles as he tells us about the record 80-kilo tuna he once caught. Then glumly he repeats that the government's fishing agreements with foreign nations are "bad business". Manuel, his younger brother by three years, has also left his own boat, he tells us. Now he works on board a big Spanish fishing vessel.

"They stay out to sea for 60 days. They never visit the harbour to land their catches; they tranship all the fish at sea. After two months they spend five days in port stocking up and then leave for another two months."

Carlo falls into silence. Then he asks for a piece of *snus*, the oral tobacco popular in Sweden, and someone gives it to him. Probably encouraged by my very empathetic nodding at this point, his story suddenly takes another, completely unexpected, turn.

"I'd like to ask you something," he says. "I need advice. I don't know what to do. My wife ..."

His eyes suddenly well with tears.

"I don't want a divorce but she's met someone else. He's been in our house and the whole village knows. Everyone. It makes me so ashamed."

They have a five-year-old daughter, he says; Maria. Carlo had named his fishing boat after her; the one he had to sell a year ago. After selling the vessel, he joined the crew of a German luxury sailing boat, where his work included sleeping on board at night to deter intruders. He had been guarding the German boat in the harbour for six months when the truth about his wife reached him. Now, he says, he has lost everything. His family, his boat, his profession, his reputation, his pride.

Carlo's frankness shocks us, but we try our best to offer some supportive words, and ask questions to make him think of a way forward. Carlo listens, dries his eyes, and rejects every proposal we make the way depressed people normally do. We offer him another beer and by the time we reach dessert he is still sitting dumbly at our table as if he thought we ought to have come up with an answer to his tragedy. A better answer.

"Stay with your wife," I venture. "If you love her, put your pride to one side. Time heals all wounds."

He looks sceptical, and I feel a sudden weight over my chest. I do not want to exaggerate matters, but cannot help myself. If anyone understands the link between his lost pride and the European fisheries policy that my taxes help to pay for, then I do. If anyone understands the link between his tears and the cheap tins of tuna at my local supermarket, I do.

He finishes his beer and finally rises to his feet with noticeable reluctance, and leaves.

Later that evening we see him standing, his head hanging down, outside one of the town's more expensive hotels. Still apparently searching for a solution. Still believing that the answer to his problems is with us, the Westerners.

Fish consumption has almost doubled in the world since the early 1970s. Not merely because there are more people on the planet; but also because people eat more fish per capita. In 1973 the world average was 11.6 kilos of fish per year; by 2003 the figure was 20 kilos. The developed world is eating the most. Average consumption in the EU is 22.7 kilos, and Japan leads the score with over 70 kilos per person per year. Developing countries – home to the lion's share of the world's fish resources – are annually consuming (excluding China) a modest 9.2 kilos per person.

Fish is in fact one of the world's most important export commodities. More than 40 per cent of world fish production is marketed internationally and trade flows are primarily going from south to north. The EU, for example, can meet only one third of its fish demand from its own waters, despite being the world's third largest fishing nation – after China and Peru. Two thirds of the fish eaten in Europe are imported; and some of the "domestic" production is in reality fished by European trawlers overseas, particularly in the developing world.

The EU's international fishing agreements originally emanate from the situation that occurred when the United Nations Convention on the Law of the Sea (UNCLOS) was concluded in 1982. Until then, the fishing fleets from industrialised countries had virtually unhindered access to the world's oceans, but this changed when the convention introduced exclusive economic zones of 200 nautical miles for all coastal states. At this point industrialised nations were forced to enter bilateral agreements with coastal states to ensure continued access to foreign fishing grounds. The EU signed a string of agreements with northern states like Norway, Iceland, Poland and Russia that effectively involved swapping quotas (some of these no longer apply for various reasons), and also with southern countries including Angola, Equatorial Guinea, Côte d'Ivoire, Gabon, Gambia, Guinea, Guinea-Bissau, Cape Verde, Kiribati, the Comoros, Madagascar, Morocco, Mauritania, Mauritius, Mozambique, São Tomé and Príncipe, Senegal, the Salomon Islands and the Seychelles. Nineteen of these countries – 15 of them from the South – had protocols in force with the EU in 2011.

When Spain and Portugal joined the EU in 1986, the cost of fisheries agreements with southern states skyrocketed for the union. These two fishing countries, particularly Spain, already had extensive bilateral arrangements with numerous coastal countries prior to EU entry and made sure to retain these when joining the union. This is why the EU Common Fisheries Policy's so-called "external policy" seems so lopsided in favour of the Spanish.

In 2005, twenty per cent, or €200 million, of the EU fisheries budget was spent on fisheries agreements (FPAs) that were 60 per cent used by the Spanish fleet. The Spanish benefited even more when you look at the value of these catches, taking home no less than 82 per cent of the €485 million landing value in that year. France and Portugal were the next most adept at exploiting these agreements, each accounting for around seven per cent of landing value. Italy and Greece had around one per cent each, followed by the UK and Ireland with less than one per cent together. (By 2009, the EU's expenditure on these agreements had decreased to €160 million.)

Debate about the agreements has been heated at times, but the EU maintains that they are better than nothing ("if the EU doesn't buy access to these fishing grounds, some other less responsible countries will"), and that they can be improved to provide more benefits to the so-called "partner countries". The basic philosophy is that the southern countries themselves lack the means to catch the fish caught by the EU fleet and hence the Union's boats are simply profiting from a "surplus" that would otherwise not be utilised by humans (fish that basically would end up dying of old age, or as food for other marine creatures), and are thereby contributing positively to the economies of the countries concerned.

EU states also claim the agreements lead to job creation, knowledge transfer and increased trade with the partner countries; thus a positive flow to the developing world. But studies on this issue, collated in a 2005 report by Swedish researcher Mikael Cullberg, show that the benefits which flow *to* the EU are much greater. First of all, the EU pays only €200 million for catches that have an ex-vessel value of €485 million or more, i.e. before value is added through processing. Secondly, the agreements create additional incidental value for the EU of €650 million, most of which arises under those with countries in the South. The deals also guarantee the employment of 13,440 EU fishing boat crew and a further 20,000 jobs on land.

And jobs created in the partner countries? According to evaluations only 2,400 crew members from third countries have obtained jobs on EU fishing boats and only around 5,000 indirect job opportunities have been created (note that these jobs on land would probably survive even if the agreements were terminated since processing the fish occur, whoever catches them).

The 2,400 jobs created are better than nothing, of course. But on the other hand, how many small-scale local fishermen, fish processors and fish merchants have been *deprived* of their incomes in the partner countries, because of foreign fisheries agreements? In West Africa many claim that fishermen for instance in Mauritania, Guinea, Guinea-Bissau and Senegal have suffered severely because of overfishing caused by foreign fleets – not only European fleets of course, but also Chinese, Korean, Russian or Taiwanese. No official estimates have been done, but no doubt the risk of overfishing threatens millions of jobs across Africa. In Senegal alone fishing employs 600,000 people and accounts for the majority of jobs in coastal communities. In 2006, Senegal terminated its longstanding fisheries agreement with the EU, in fact the oldest one, originating in 1982. The reason was that fish stocks were dwindling, and no longer abundant enough to meet both domestic and foreign demand. Unfortunately, this did not hinder European operators to go on fishing in Senegalese waters – under new flags, and private agreements.

So what about positive spin-offs in trade? The theory is that fishery agreements should lead to higher trade flows for partner countries, but again the reality is different and partner countries receive little benefit. According to the French maritime research institute Ifremer, the EU long-distance fleet makes three quarters of its purchases of goods and services from companies in Europe (€154 million) and just one quarter (€51 million) from partner states. Ifremer's data also show that the value added for partner countries is substantially lower than for the EU. A number of studies performed by independent researchers to provide data to the European Commission confirm that the developing states hardly ever even see the European vessels in their harbours. Even if most agreements stipulate that the EU fleet should land part of their catches in the partner country, employ local observers on board, and regularly report their actions to the third country's authorities – this is not happening. Normally the EU fleet catches the fish in West Africa, and lands it in the Canary Islands – not contributing at all to trade or local economies in Africa.

So what about the fundamental principle of the so-called Fisheries Partnerships agreements (FPAs) that the EU fleet only fishes the partner country's surplus, harvesting a resource that would not be utilised otherwise?

The situation is far from clear according to two studies in this area, one by the French Marine research Institute Ifremer, and one from the Belgian consultancy Aide à la Décision Économique, commissioned by the European Commission's aid agency, EuropeAid. Both show that agreements are based on shaky data: stock assessments are uncertain, social impact assessments not carried out, monitoring and control glaringly absent. Adding to uncertainties is the fact that a majority of the agreements are with countries prone to corruption and with weak democratic institutions, which means that if a regime guarantees that it is only selling a "surplus" it doesn't mean the local fishing sector, or independent researchers, would agree. In many cases it is evident that the EU fleet is competing with local fishermen for the same fish stocks. And sometimes the EU fishing methods – for instance for shrimp trawling – are appalling. 90 – 95 per cent of the trawl catches are unwanted, so-called by-catch, and are discarded. So – even if the targeted shrimp is considered a "surplus" – the by-catch is not included in that estimate, and might be either threatened species or perfectly edible species that are normally consumed locally.

With regard to the high level of corruption in some countries, there is a very telling story, confirmed by several sources, about a former fisheries minister in Guinea Conakry who openly tried to bribe an official from the European Commission, giving him an envelope containing money, as "a gift to your children". In other words it is not always the case that the EU is putting pressure on partner countries to sell as many fishing opportunities as possible, it might well be the other way around. In this case Guinea wanted to sell as much as possible to the EU, in order to receive as much money as possible, but the EU didn't believe there was that much fish left to sell. The EU official turned down the offer.

Anyway, the fact remains that a number of the EU fisheries partner countries fare extremely badly in Transparency International's rankings of the world's most corrupt nations. The index of corruption among politicians and civil servants rates from 10 to 0, where 0 means high and 10 low corruption. The "partner countries" Guinea Conakry, Guinea-Bissau, Côte d'Ivoire, Senegal and Mauritania all rate between 2.0 and 2.9 on the index, which means very high risks of corruption taking place.

On the issue of "surplus" it is also highly dubious to talk of a surplus of fish in countries like Mauritania or Guinea-Bissau, where EU economic payments for fisheries access have accounted for up to a third of the countries' budgets. If these countries focused on developing their own fisheries sector, like Namibia did when it gained independence in 1990, they would doubtless earn much more than they do from allowing foreign vessels into their waters.

The Namibia case is a good illustration of the benefits that countries can gain from preserving their fishing autonomy. Conditions for productive fishing are perfect off the Namibian and Angolan coasts thanks to the Benguela current upwelling of nutritional waters. When Namibia was governed by South Africa in the 1970s and 1980s, fishing fleets from many countries, primarily the Soviet Union and Spain, flocked to the country's poorly controlled waters. Hake and sardine stocks were overfished to critical levels. Foreign trawlers accounted for between 90 and 99 per cent of hake catches when Namibia became independent in 1990 – and sardine stocks were just two per cent of what they once were. Annual per capita fish consumption in the country was just four kilos on average.

The new Namibian government moved quickly to asset its exclusive economic zone of 200 nautical miles, resisting pressure to sign a new fisheries agreement with the EU. It closed the hake fishery to foreigners and set a conservative quota of 60,000 tonnes. The EU questioned this quota in an attempt to win an agreement to fish 200,000 tonnes; Namibia responded by offering an allocation of 9,000 tonnes – on the grounds that this was the only "surplus" beyond its national needs. Brussels pulled out at this point and there have been no bilateral agreements since.

Namibia's own fishing industry has since gone from strength to strength. Domestic companies and Namibian-controlled joint ventures have priority to all fishing rights ahead of foreign fishermen. Fishing rights are allocated as seven-year quotas for foreign-controlled companies and ten-year quotas for Namibian firms. Quotas can be reviewed, but not transferred or sold to third parties. Quota holders are required to pay an annual fee to the state and a smaller levy per tonne of fish landed, and these revenues are used to finance the Ministry of Fisheries' budget. Fishing boats are required to foot the bill for having observers on board to control their catches and collect scientific data. Observers are assigned by the ministry and spend limited periods on each boat to avoid any risk of dependency on the ship owner. Coastguards patrol the exclusive economic zone and illegal fishing has become less of a problem following a number of arrests, with monitoring now focused on ensuring that legal vessels comply with government regulations and quotas.

And Namibia's fish stocks have partially recovered – and would, according to experts, have been at least 50 per cent more profitable than they are today had it not been for the foreign trawlers that plundered the fishing grounds prior to independence. All the same, it has been a success story, and although recent jellyfish blooms are probably signs of overfishing, fisheries have become an economic engine – Namibia's fastest growing industry after tourism and the second largest export earner after mining. Before independence, fisheries accounted for just four per cent of GDP; now the figure is ten per cent. The number of people employed in the sector has more than doubled from 6,000 to more than 15,000. Two thirds of them work in the fish processing industry.

Domestic fish consumption has also more than doubled, despite the fact that Namibia exports 95 per cent of its fish, primarily to the EU and southern African states. The government has also introduced exemptions from quota fees for fish species traditionally popular in Namibia, provided they are landed domestically. This has led to Namibians' fish consumption increasing to the same average per capita level as the rest of the developing world – around nine kilos per year.

Neighbouring Angola took a leaf out of Namibia's book in 2004 by introducing a fish resource reform that imposed tough new requirements on EU vessels – so tough that Brussels declined to renew the bilateral partner agreement which had been in place since 1989. Among the new stipulations were that all fishing boats should be connected to an Angolan company and that all foreign vessels should be equipped with a VMS monitoring system to facilitate tracking by the Angolan authorities. This was an offer that was turned down by the EU, fish surplus or not.

Another tricky issue when it comes to stipulating a proven surplus of fish, is the accounting for highly migratory species, such as tuna. To what country does a "surplus" of fish really belong, when the fish move around, crossing national borders all the time? Today this fish "belongs" to those nations that have fished the most in the past, establishing "historical catches". In ICCAT (International Commission for the Conservation of Atlantic Tuna) and other regional fisheries management organisations where tuna quotas are negotiated internationally, the industrialised world has taken the largest share of the world's tuna and swordfish quotas, by showing their historical right to that fish – simply by proving they used to fish them in the past. But another allocation system could be possible – and is suggested by many. What if the amount of migratory fish such as tuna and swordfish caught in a country's EEZ, belonged to that country and not to the country that used to fish them? In a fair new world, coastal states in the developing world should be allocated their own quotas of tuna stocks which they at present don't have since they have never fished them in significant numbers before. But what if they had the means to fish the fish themselves, and sell the fish with added value themselves on the world market? Tuna agreements are indeed "positive" for the EU, no one disputes this. So, who honestly thinks that given the option of exploiting the resource themselves, developing states would generously sell off access to their waters, as if it were just a useless "surplus"?

Finally – how about the argument that bilateral fisheries agreements contribute to establishing fisheries policies, and strengthen the local fisheries sector in developing countries? Since 2002 all EU bilateral fisheries agreements with the South (so-called Fisheries Partnership Agreements) contain a clause earmarking funds for national fisheries policies, research and development – and the amounts involved are increasing. The protocol with Cape Verde stipulates that the entire contract sum is to be used for measures beneficial to local fisheries, including monitoring, research and development.

Nobody would deny that the countries involved need support for fisheries management, research, or control of their economic zones. But in the Cape Verde case, according to independent evaluations performed by consultants to the EU Commission (in 2011), monitoring, control and surveillance systems have been inexistent practically throughout the duration of the protocol, and still there is no national fisheries policy in place.

A Greenpeace action off the West African coast in 2006 revealed the true extent of illegal fishing: in three weeks spent patrolling Guinea Conakry's economic zone, the Greenpeace ship *Esperanza* spotted 104 fishing boats, half of which were suspected of illicit fishing. Some had no licences, some were fishing inside the 12-mile zone reserved for small-scale coastal fishermen, some were unnamed pirate ships, and others concealed their identity from the Greenpeace crew. Many were suspected of transferring catches to other vessels at sea, a practice officially banned in Guinea because landings are the only way to monitor catches.

A European contribution to improved monitoring is of course more than welcome, but "partnership agreements" that stipulate the use of the money received by one "partner" raise some fundamental issues. "The EU's desire to control the use of funding can be viewed as a paternalistic deviation in a commercial agreement between sovereign states," as Susanna Hughes puts it in her 2004 report *Effects of EU's Agreements on Fishery in Developing Countries*. According to her the fisheries agreements are dubious from a development aid perspective because developing countries only receive support to control their waters if they at the same time sell their fish to the EU. As in Senegal's case, the lapsed agreement with the EU cut the funding to the country's fisheries monitoring administration: when the fish are gone the "aid" flow stops.

On the other hand one might hope that the earmarked appropriations prevent fishery funding from ending up in the pockets of corrupt officials, or being used for other purposes. But, unfortunately, so far lax monitoring of the use of the funds provided by the EU fisheries agreements has meant there is little control over where the money ends up. Very little has improved under the FPAs, according to numerous evaluation reports paid for by the European Commission. Partly because the EU has not done proper follow-ups on how the appropriations have been used, partly because many countries, like Mauritania, Angola and Senegal, have actually opposed making binding undertakings to use funds in specific ways – in order to preserve their freedom to make their own budget priorities. This is an entirely reasonable approach, according to Swedish researcher Mikael Cullberg in his report *EU Fisheries Agreements in Developing Countries*. "It may be in order to point out that most states, including Sweden, have a fundamental principle that government income should not be tied to specific forms of use."

Another point of principle is that the EU's heavily subsidised fleet competes with domestic, unsubsidised fleets on export markets creating a non-level playing field. This was why Morocco, to the dismay of Spain, terminated its fisheries agreement with the EU in 1999. Moroccan fishermen, fishing alongside their EU counterparts in their own waters found it impossible to match the price of octopus sold by the EU boats to the Japanese market. Mauritania, another country rich in octopus, has the same problem. Its fishermen pay twice the price for their diesel as EU fishermen do thanks to subsidies from Brussels, forcing them to charge at least 30 per cent more for the their own octopus on the export market. Unfair competition indeed.

Another potential problem is that domestic fishermen are encouraged to sell their fish for export rather than for local consumption when new sales channels are established in developing countries. The Cotonou Agreement of 2000 allows all countries that has fisheries agreements with Brussels (if they have approved phytosanitary conditions, which is not always the case), to export duty-free to the EU. An advantage that has encouraged the African, Caribbean and Pacific Group of States (ACP) to focus on European exports. Some 63 per cent of their fish exports are to the EU, with 27 per cent going to Japan and 10 per cent to the US. Competition from attractive export markets can drive up prices in local markets or result in the entire catch of fish species popular in the West being sold abroad rather than locally. In Senegal, the bottom-living fish that used to be bountiful are now hard to buy locally because they fetch higher prices in Western markets. Senegalese consumers must instead rely on less valuable, smaller pelagic species.

A number of reports suggest the consequences of overfishing by both domestic and foreign fishermen in Africa are also taking another, even more worrying turn. As industrialised countries have laid claim to large parts of developing countries' fish resources, so hunting pressure on terrestrial animals in Africa has increased in turn. A conference in Madagascar in June 2006 confirmed that pressure on wild animals in Africa had become alarmingly high, with "bushmeat" (hippopotamus, antelope, chimpanzee, elephant, gorilla and others) being an increasingly popular and cheap source of animal protein for Africa's poor. Researchers in Ghana, the UK and the US claim the suggested link between reduced availability of cheap marine protein in the seas and increased bushmeat hunts in nature reserves is real.

The EU is a schizophrenic institution: what it gives with one hand it takes back with the other, pumping millions of euros into agricultural subsidies to EU tobacco farmers while investing many thousands in anti-smoking campaigns. And this is just one example. Various fundamental objectives collide when it comes to EU fisheries agreements, for instance the ambition of supporting European fishing communities and guaranteeing European food security compared to the objectives of protecting the marine environment, fighting poverty, and showing solidarity with developing countries. There are numerous European and international agreements and treaties declaring that the fight against poverty should take precedence over everything else and should always be considered in agricultural, trade and fisheries policies. The "European Consensus for Development" established that non-development policies should assist developing countries efforts' in achieving the UN Millennium Development goals (eradicating poverty and hunger). Under the 2009 Lisbon Treaty it is even a legal obligation. Article 208 states that "Union development cooperation policy shall have as its primary objective the reduction, and the eradication, of poverty. The union shall take account of the objectives of development cooperation in the policies that it implements which are likely to affect developing countries."

But how do these noble ideals work in practice? The European Commission's own independent consultant, Aide à la Décision Économique (ADE) in 2001 tried to answer this question in the context of fisheries agreements and delivered a damning verdict.

ADE looked at: 1) whether the agreements were consistent with overall policy goals, 2) how the agreements contributed to poverty reduction, 3) the impact of the agreements on food security, 4) how the agreements promoted sustainable fisheries, and 5) the sustainability of the agreements themselves. The EU failed on all points.

The main shortcomings were inadequate monitoring of the agreements which led to a lack of data to provide accurate support for taking informed decisions. ADE also concluded that the agreements were not credible in purporting to be based on surplus fish stocks, noting that surpluses are extremely difficult to calculate, due to a lack of information on real fishing pressure in the zones in question. Moreover, the EU system of advance payments for a specified tonnage encourages partner countries to inflate the quotas they want to sell, which can lead to over-fishing. This is because partner countries, according to the agreements, do not have to reimburse Brussels if the EU fleet fails to fill its quota, but must do so if they themselves reduce the quota due, for example, to stock depletion. The fact that partner countries almost always sell more fish than there is to go round (a phenomenon called "paper fish") is demonstrated by the fact that the EU rarely fills its paid quota.

In other words, the EU fisheries agreements in their present state are a long way from ending up in the history books as models of sustainability, rather – of course in good company with agreements of other long-distance fishing nations such as China, South-Chorea and Taiwan – as flagrant examples of what money, bureaucracy and corruption on a grand scale can achieve together.

But – one obvious reason for keeping the agreements, and trying to improve them, exists though. Every expert agrees, including development NGOs, that terminating the agreements would not mean that EU operators scrap their boats and go home. In the interest of transparency and good governance the EU must maintain control over its own nationals, making sure they fish legally and according to principles of sustainability wherever they fish in the world. And fisheries agreements can be an important framework for EU engagement in this field. Still, no one doubts that lots of work remains before they really live up to the proud declarations in the texts of the protocols.

We are bouncing along the cobblestones through the flat, barren landscape leading to the fishing hamlet of Baia das Gatas, thirty minutes east of Mindelo on the island of São Vicente in Cape Verde.

A single street of humble, whitewashed houses fringes a paradisal beach at which foreign tour operators have been casting covetous glances for some time now. Children and dogs frolic on the tideline and the turquoise sea is calm in the sheltered bay. The children look for shells, then immediately smile at us and want us to take their photograph. They then thrust out their hands just as quickly to ask for money.

Rows of wooden fishing boats painted in typically bright Cape Verdean colours are lying with their keels in the air on the beach beside the ghost of an old fish processing plant. The building lacks roof and doors and has the word RUINAS daubed in uneven letters in red on the walls. It is around noon and most of the boats have just come in. But fishing luck seems to have been minimal today. The twenty or so men gathered on the beach are all silently gazing out to sea, pinning their hopes on the last two boats that now are close to the beach, reefing their sails.

Well, "sail" is of course a relative concept. "Look," says our German guide, Janine Hetzel, who lives on the island and speaks fluently the local Cape Verdean *crioulo*, a mixture of Portuguese and African languages.

She points at the simple wooden masts with white sail wrapped round them like scrolls, lying beside the boats. It looks as if the sails were made of plastic. Here and there I see writing on them. I look closer. "Arroz"?

"They're made of old rice sacks," she confirms. "They last a few weeks and then the fishermen make new ones. Most of the men can't afford outboard motors; *this is all very small-scale,*" she says with the understatement of the day, gesturing towards the thin men, the blown-out old plant, and the lingering stray dogs.

I nod. Small-scale, *indeed.*

Now the men on the boats jump into the sea and a dozen or so from the beach wade out to help bring the boats in. The sails are rolled up, the boats pulled up far on to the beach to be safe from the tide, and I stand by watching, really curious to see what they have caught.

Janine interprets my question, and one of the fishermen, 30-year-old Manuel Rodriges Lopez smiles awkwardly and shows me what he has got in the bottom of a bucket. Two fish. One looks a bit like a bass and is barely 20 centimetres long; the other is a metre-long silvery eel-like creature that resembles a very slender garfish but with a smaller jaw. Holding his modest quarry between thumb and forefinger with a stoical smile, he presupposes I want to take a picture of it. I do.

Evidently the eight men in the two boats have been out fishing since five in the morning but this is all they have to show for their efforts. Manuel says there is nothing unusual about this: fishing has been bad in recent years.

"The Spanish longliners are everywhere out there. They're much bigger than we are. We don't have a chance to catch anything."

While we talk, and Janine translates, all the fishermen on the beach gather round us very closely. When I ask Manuel what he thinks about other countries being allowed to fish in Cape Verdean waters and Jasmine translates my question into *crioulo,* the men burst out laughing. Manuel smiles awkwardly again and shrugs. Jasmine interprets:

"They don't like all the illegal fishing, that's all they want to say. Someone should be policing all the illegal boats around here."

A short discussion between the men now follows, and then Manuel gestures towards his boat, painted in clear yellow, red, and green colours, and explains that a boat like it costs 200,000 Cape Verdean escudos. An outboard motor, necessary nowadays to pass the neighbouring island of Santa Luzia to have a chance of getting any fish, costs 120,000 escudos. Translated into European money this is about 3,000 euros – a fortune in a country where a hotel receptionist's wage is about 100 euros per month. Working to save to buy a boat is virtually impossible, Manuel says.

I ask how much he has earned today. Manuel points at his two skinny fish in the bucket, and then at his seven fishing partners. Janine explains:

"The boat owner gets half of the catch and the crew share the rest. That's the usual system here."

An older man with silvery stubble on his dark complexion now penetrates the crowd to show me the contents of a sandy, semi-transparent plastic bag. I look inside and an aroma of sea strikes me, emanating from sea shells of different forms lying in there. "He wants to sell them," Janine says.

"And considering all the information the men have shared, I think it might be a good idea to pay him," she advises.

I rummage in my pockets and hand the man a few hundred escudos. With a smile, he hands over the plastic bag. When we are about to leave, I notice that Manuel, who has spent the most time talking to me, looks disappointed. And I understand why. Of all the people on the beach today, the man with the bag of shells made the biggest profit.

In the car, on our way back, we pass newly built bungalows and Janine tells us about the annual music festival in Baia das Gatas that draws thousands of visitors. I sit thinking of the men on the beach. That today they probably experienced something they already knew all too well: that their fishing village is changing, slowly but surely. That we Europeans home in on them from two directions. First we have taken their fish. Now we are going to have holidays – in their fishing villages.

THE EUROPEAN UNION AND THE DUTY TO EXPLOIT

Brussels, October 4, 2004. The Altiero Spinelli Building[6].

I tingle when I spot the sign. For months I have been trawling the Internet for documents that could explain this or that, reading innumerable resolutions, reports and communications from the European Parliament, the Council or the Commission, and each time I have seen these words in that typically bland EU typesetting: "Commission de la Pêche – Committee on Fisheries".

And now, there they are, the words in black and white – on a sign on a black double door to Room A1E2, 1st floor in section E in one of the European Parliament's big glass buildings in Brussels, exactly as announced. I almost chuckle. What has nearly become a fictitious entity, that intangible power in Brussels, the "Brussels" everyone talks about, from Smögen to Cape Verde – actually exists. Behind those doors, in just a few minutes, an ordinary run-of-the-mill meeting will take place and I am holding the agenda in my hand. The main item for this meeting is the upcoming hearing with the new Commissioner for Fisheries and Maritime Affairs, Joe Borg, a Maltese who has been nominated to succeed the Austrian Franz Fischler who has held the position for the last few years.

The commissioners are the EU Commission's "cabinet ministers", appointed for five years. The "prime minister", European Commission President José Manuel Barroso, has already been approved by the

---

6   Please note that the author was not a member of the European Parliament when this chapter (or indeed book) was written.

members of the European parliament (MEPs), after a hearing in the summer.

The hearings with the other commissioners are held over little more than a week, and this time they prove to be more than a mere formality: Italy's nominee for Commissioner for Justice, Freedom and Security, Rocco Buttiglione, a Christian Democrat, makes a huge gaffe with his views on homosexuals and women, and is rejected by the MEPs – a unique EU moment that shifts power decisively in the direction of the Parliament.

Probably, to most European citizens, the EU institutions and corridors of power in Brussels and Strasbourg, are a quite a mystery: the Commission, the Parliament, the Council of Ministers, the various directorates and committees ... Who decides over whom is very unclear in a European context, and I am just starting to learn.

The Committee on Fisheries, or "Commission de la Pêche", as it says on the door, consists of elected MEPs who have opted to work with fishery issues. For the 2004-2009 period they number 55 (including 22 substitute members), mostly from Spain, the UK, France, Italy and Portugal. Most have their constituencies in their countries' coastal regions, which basically means they see as their task as protecting the interests of voters in these areas.

The Parliament's Committee on Fisheries is in other words often considered as the fishing industry's extended arm in the EU; a very conservative force that for decades has critically examined and opposed almost all progressive proposals coming from the Commission.

Sadly enough, both the Green Paper on the reform of the Common Fisheries Policy of 2002 and the proposal for new objectives for the 2007–2013 programming period were met with indignant storms of questioning and angry statements from the members of the committee – who in fact are the only *directly* elected EU representatives that can sway fishery policies (even though the European Parliament had only an advisory role in fishery issues before the Lisbon Treaty took effect in 2009). Apart from this, EU citizens who want to influence fisheries policies on election day have to rely completely on the proportional vote in relation to population that their ministers for fishing exercise in the Council of Ministers; in Sweden's case, a vote that has been constantly voted down.

So what about Sweden's representation on the parliament's committee? Well, there is one Swede; although he is only a substitute member. He is Carl Schlyter, a Green Party member born in 1968 in the Stockholm suburb of Danderyd, and one of four committee members belonging to a group called The Greens/EFA (European Free Alliance). The committee as a whole is largely dominated by the Christian Democrat group EPP-DE (European People's Party – European Democrats), the ALDE group (Alliance of Liberals and Democrats for Europe) and the S&D group (Progressive Alliance of Socialists & Democrats).

But what role does the European Parliament's fisheries committee actually play in the decision-making process? If Swedish decision-making is lengthy – from researchers' observations via ICES, the Board of Fisheries, the Ministry of Agriculture and the political authorities, before reaching the Council of Ministers and the EU – *within* the EU it is almost endless, and typically EU bureaucratically complex and labyrinthine to boot.

The Council of Ministers (the fisheries ministers in all member states) asks the Commission (roughly, the EU equivalent of a government cabinet) to put forward a proposal; or it could also receive a proposal from the Commission, prepared by the Directorate-General for Maritime Affairs and Fisheries (previously DG Fish, now DG Mare). DG Mare constitutes a large organisation that takes every conceivable aspect of fisheries into account across various sub-sections. If the issue is, say, curbs on fishing with potential consequences for regional policy or the labour market, other Directorates-General are involved, along with committees of representatives from member states and industry or scientific experts. One of these is the Advisory Committee on Fisheries and Aquaculture (ACFA), established in 1971 and with representatives from across the fishing industry and also from NGOs and consumer groups.

Proposals prepared at this level are then sent to the Council of Ministers, the decision-making body for the member states (after the Lisbon Treaty came into force in 2009, proposals also have to go to the European Parliament for its endorsement, so-called co-decision). No matter how wellprepared a proposal is, it can still be torn up at this new level, since this is where member states are directly involved, represented by their fisheries ministers, and can say yay or nay to what would become Community Law. The Council has its own groups for internal and external fisheries policies who seek to find a balance between the different positions of the member states. At the same time, the proposals will have gone to the European Parliament's Committee on Fisheries for deliberation.

Then (are you still with me?!) when committees and the Parliament have had their say, the proposal goes back to the Council of Ministers, is adjusted a notch by the Committee of Permanent Representatives, after which the fisheries ministers vote. Decisions are reached by a qualified majority with member states' votes weighted according to population. For many years, Sweden has voted in vain against cod quotas that are out of step with scientific recommendations. The notorious consideration given to "socio-economic aspects" half a dozen times or more only at the Swedish level, surfaces again at least as many times in Brussels, which offers the only reasonable explanation for the blatant ignoring of scientific advice. Instead quotas have been inflated to the point that Baltic fishermen for a number of years didn't even manage to catch the whole quota, despite considerable effort.

Anonymous women in black skirts and high-heels swish by, to disappear through the black doors, leaving trails of perfume. They remind me of receptionists in continental hotels; by the way, the entire atmosphere is a little like that of a 5-star hotel. Light grey carpets absorb all sound, only the ping of lift doors regularly opening and closing nearby breaks the silence. A faint smell of cigarettes is in the air, and a distant hum in French and German.

"Madame, madame!" a polite woman says when I attempt to enter the committee room at the appointed time for its meeting. "The first half-hour is for coordinators only. You can come back at three thirty."

I go back to sit in the austere, black leather furniture group outside the committee room and watch the "coordinators" – a few casually dressed men and women – stroll in through the swing doors. A couple of thin young men in suits rush out; perhaps they too have thought they could attend the meeting from the appointed hour.

Now the women in dark clothes and clinking accessories exit again.

As I suspected, they are some kind of hostesses. Speaking French and Italian, they settle behind a small counter of expensive red wood where they shuffle piles of papers in various translations. A small delegation of waiters in white shirts and black bow ties suddenly appears with a silver coffee trolley. Everybody is very discreet, refined and somehow dignified, and I'm beginning to grasp the special "European" atmosphere that sooner or later seems to absorb everyone who ends up in Brussels. Already I too feel a bit infected by it – the important hastening along soft carpets and the conversing in various tongues – it is inspiring and enhancing, a breath of something important and international that ought to be good. Still, I can't help but grin when I see the words European Parliament in 15 spellings on every sign and hear bartenders in the large members' bar prudently taking orders for coffee – *café con leche, café au lait* or *caffelatte* – as though the right of Europeans to have their own variety of coffee with milk was a major issue.

Finally it is half past three and the elegant hostesses at the counter rise and open the doors for the large number of people now waiting outside. The committee will have nine meetings this term, eight of them in two successive days to make it more convenient for members to attend. I take a seat at the back of the large conference room, on one of about ten seats marked "Presse". I am the only journalist here.

That I am the only representative of the Fourth Estate does not in the least surprise me. Already I am aware that there are scarcely any Swedish journalists specialising in fishery issues; it seems to hold also for my European colleagues.

The reason for the lack of reporting on fisheries is of course the subject's almost endless complexity – and its invisibility. True, people come into contact with fish daily: we see fish fingers in the supermarket freezer, fish on the restaurant menu, eat-more-fish exhortations from health gurus and, in some countries, there is a growing debate about which fish you can eat with a good conscience. But doing in-depth reporting on fisheries policies or science is another matter, and what journalist has the time to digest the humungous volume of information and counter-information that the authorities, professional fishermen, NGOs, scientists and the EU are constantly churning out? The technical language in these documents turns a news reporter off already after the first sentence. What are pelagic species? Demersal? What is meant by ecosystem approach, recruitment damage and purse-seining? What is the RAC, the TAC, the Blim, and the BFT? What is the difference between "red tuna" (*thon rouge*) and bluefin tuna, and why do both seem to have yellow fins on the pictures? A reporter, with a deadline within a few hours and an editor in charge who won't give him or her more time to do research, has only one possible recourse: to call an expert. Then, to make the story as non-partisan as possible, the reporter tries to find someone who does not agree with what the expert is saying, and we all recognise the result and could probably do the job ourselves in our sleep: the news clip with a scientist talking about impending catastrophe; then a voice-over with the reporter saying something like "but the professional fishermen don't agree", then cut to a professional fisherman who (you've guessed it!) – doesn't agree. One minute and twenty seconds, tops, and then the news moves on to more prestigious stories, like those on politics or economy.

Anyway, at least now I'm here, attending the Committee on Fisheries meeting and contemplating that the EU's ocean region occupies more area than its terrestrial surface and that soon, as the union grows to include 27 members, even more inland lakes and sea areas will come under EU jurisdiction: the Black Sea, the Mediterranean, the Baltic, the North Sea and the Atlantic – and that the EU countries also are major fishing nations in the rest of the world's seas. The wellbeing of fish in these seas influences the entire marine ecosystem down to the level of the plankton. And the seas, in turn, influence us humans. Directly, since they provide us with food and clear waters to swim in and indirectly because the seas, covering the major part of the planet's surface, are a crucial climate regulator. So – if fish stocks collapse, the marine ecosystem collapses, and climate change accelerates out of our control.

So – where are all the reporters, I wonder?

I put on headphones and twist a dial beside my seat and hear in language after language what the chair of the Committee on Fisheries, former French military general Philippe Morillon, is saying. Behind the smoked glass of booths above the small plenary room, simultaneous interpreters from every country in the Union, impressively versed in the notoriously difficult fishing terminology, are talking quicker than I can write. Morillon is presenting the EU's programme for 2007–2013 and the new European Fisheries Fund (EFF), totalling about €7 billion, which will replace the earlier Financial Instrument for Fisheries Guidance (FIFG). The Commission's proposal, which calls for a further reduction in fishing effort and more protection for the environment, provokes the old gentleman, member of the French *legion d'honneur* to exclaim that he is "shocked".

"It seems strange to me to protect fish if there's no one to fish them. We have a legitimate goal to protect natural resources of course, but nowhere in this proposal is there mention of the industry's situation," Morillon intones.

Paolo Casaca, a socialist from Portugal, objects saying that fish must be the first priority since if there is no fish there will be nothing to catch, and a blonde Spanish member from the Conservatives in her fifties, Carmen Fraga Estévez, says the proposal makes her "sad" and that she finds it "unacceptable if EFF money cannot be spent on the modernisation of the fleet".

Between 2000 and 2006, around €800 million (of which €300 million were national funds) were spent on the scrapping of vessels, while another €800 million was awarded to the construction of new vessels and €400 million to modernising existing vessels. €450 million went to aquaculture (fish farming), €375 million to harbour infrastructure, €950 million to marketing and the processing industry and almost €200 million to socio-economic measures. On top of this, more than €2 billion were spent on "other" purposes, mainly the acquisition of fishing rights in foreign waters in the developing world.

The EU is not alone in subsidising its fishing fleet. According to the most recent figures (UN Environment Programme, UNEP, 2011), the world's fishing fleets get a whopping USD 27 billion in support each year. Japan leads the league, but the EU is second by granting at least 3 billion USD a year to its fishermen.

The original purpose of all this tax money was of course to give the world access to good quality fish, cheaply. But the result has been the opposite: for a continuously diminishing supply, taxpayers are paying good money twice or three times, first through income tax to subsidise the industry, then in the supermarket where fish like cod that was once cheap has become expensive, and often a third time, when professional fisherman deprived of catch because of fish depletion, need income support. The Canadian story is infamous. Less well known is that EU taxpayers have compensated Spanish and Portuguese fishermen to the tune of €197 million because the EU was unable to renew the fisheries agreement with Morocco in 1999. And fishermen who "suffered" when the EU fisheries agreement with Angola was not renewed could also apply for compensation.

(Talking about losing public money: the UN report "Towards a Green Economy" shows that in a "green scenario", where the world's fish resources were managed in a responsible way, the incomes from fisheries globally would increase from USD 17 billion a year to USD 67 billion a year by the year 2050. Each year with mismanaged fisheries the world loses 50 billion US dollars; "the sunken billions".)

Next point on the Fisheries Committee agenda concerns, fittingly, one of the EU's many fisheries agreements with developing countries. 38 Spanish tuna vessels, 18 French, 5 Portuguese and an Italian vessel have, with the help of €292,000 of taxpayers' money, been granted access to the waters of the Comoros, the little Islamic republic in the Indian Ocean. The white-haired committee chair Morillon runs through the protocol, most of it old, since the EU has had an agreement with the Comoros since 1988. And as usual, the Commission had already signed the protocol in February, six months previously, and not until now have the European Parliament and the Committee on Fisheries had a chance to express an opinion.

Elspeth Attwooll, a Liberal Democrat from the UK, finds the agreement completely unacceptable and suggests the Committee arrange a hearing on the hidden subsidies for shipowners. And the lone Swede in the Committee, Carl Schlyter, suddenly springs into life. Like me, he has long since removed his earphones, leaving the Swedish interpreter behind the tinted glass translating into empty space. Schlyter now questions that the Commission has signed the protocol without waiting for an evaluation of the earlier agreement or even for the scientific fish stock assessments that should be the basis of any new protocol.

As far as I understand, these objections are merely noted.

A third point on the agenda is about protecting coral reefs in the Atlantic, and now the different positions of MEPs start to become clearer. The Spanish member, Carmen Fraga Estévez, questions whether the reefs really are located where they're said to be and if bottom trawling really does cause the damage alleged. Chair Morillon wonders, with brutal directness, whether the EU really should make laws for everything, and if there really is a need for a law to protect coral reefs at all. The rapporteur (the member responsible for the coral reef report) Sérgio Ribeiro explains that the coral reefs around the Azores, Madeira and the Canary Islands were previously under national legislation and it is now the EU's duty to step in to protect them.

The result of this discussion with these different approaches and objections is not completely clear to me. My remaining impression is, anyhow, the members' general lack of prior knowledge of the issues.

The following session is at nine the next morning, this time held just for the ten committee members, some staff and me (and the numerous interpreters). The blonde Spanish member Carmen Fraga Estévez is here of course, and there is something about her strikingly self-assured attitude that draws my attention. She represents the Partido Popular and has been a European parliamentarian since 1994. She is at this point first deputy chair of the EPP-ED, the conservative Christian Democrat grouping, and was chair of the Committee on Fisheries between 1997 and 1999 (a post she would regain after the 2009 EP elections).

This morning the recovery plans for southern hake and nephrops – two stocks teetering on the edge of collapse and with radical EU measures for recovery already approved – are going to be discussed, and a sizeable delegation of Spaniards in dark suits suddenly files into the row in front of me. Cheek-schmoozing greetings break out between them and Mrs Fraga and some other Spanish MEPs. The next point on the agenda is indeed "Discussion with the fishing sector".

That I happen to witness a parliamentary meeting where the Spanish sector is invited is of course not a coincidence: Spain is the EU's predominant fishing nation, landing over 1 million tonnes of fish a year (against an EU total of 4.5 million) to a value of almost €1.9 billion (against the EU total of €6.2 billion). The Spanish fishing industry employs more than 55,000 people and has, at this point, 14,000 boats of the EU's almost 90,000, many of them huge. Spain has a quarter of the total EU tonnage, and receives almost half of the funding.

Here, obviously, are interests to protect, and the Spaniards in the row in front of me seem at home. It is not unusual for interest groups to request to address committees in this way and now it is time for an impeccably dressed, dark-haired woman to be given the floor. She is the spokesperson for a delegation of fishermen from Lugo on Spain's north-west coast. The agenda gives her name as Mercedes Rodrígues Moreda. She is in her thirties, has curly hair and says she represents 125 shipowners, 50 of whom are affected by the Commission's proposal to reduce quotas for southern hake and nephrops. She says that 1,200 jobs are threatened and that another 3,000 boats in Spain and 2,000 in Portugal would be adversely affected.

"Naturally we are not against protecting the southern hake," she says, "but some things do not reflect realities. The protective measures would affect communities socially and economically."

She pauses for a short moment to brush the hair away from her forehead, and then continues with great emphasis:

"Please excuse me if I get carried away; I get very emotional, but you cannot create poverty in our region!"

She takes a breath to calm down, and then refers to a certain doctor in marine biology who supports her view that nothing can be taken as certain.

"For 15 years now the southern hake has been below safe biological levels but even though fishing has continued nothing has actually happened. Why hasn't the stock collapsed? It is clear that the reference points are invalid and the safe biological limits are questionable. We need more science!"

She calms down again and continues:

"Naturally, preserving fish resources is good, but it is important to preserve families and fishermen too. We want to work with scientists but sometimes it feels like we are working against them. We are interested in ecology but the scientists' assessments are totally vague and uncertain!"

And then a final salvo:

"We do not accept reduced quotas! If they are lowered now, they will never be raised again. Things have got completely out of hand, *completely out of hand*!"

The Spaniards present nod in agreement and I recall reading something about how the Spanish fishing fleet once could expand so enormously. Awkwardly enough, one might say that the late Spanish dictator, General Franco, still casts his shadow into this very room in Brussels in 2004. One of Franco's ambitions was for Spain to be a world power at sea, and the fishing industry attracted his particular attention. Generous stimuli more than tripled Spain's fish catches, from 400,000 to 1.5 million tonnes annually between the end of the Civil War in 1939 and Franco's death in 1975. Ever since, in contrast to the other big fishing nations such as the UK, Spain has done everything to avoid reducing its fleet.

But just how large Franco's shadow was looming I didn't really realise until much later, when I Googled the Spanish member Carmen Fraga. It turns out she is the daughter of Manuel Fraga Iribarne, founder of the right wing Partido Popular, and before that, propaganda and tourism minister under Franco from 1962 to1969. Manuel Fraga was active in Spanish politics up until the summer of 2005 when he resigned after an election defeat. He was born, I read, in the village of Vilalba, not far from Lugo.

I go through the motions, questions and resolutions of the different members of the committee, and it doesn't take long to see that the Spanish members are the most active ones, not least Mrs Fragas' compatriot, the socialist Rosa Miguélez Ramos.

One of the most controversial debates concerns sharks. In 2003, the EU banned the cruel practice of "finning", by which the valuable fins are cut off and the sharks, still alive, are dumped back in the sea. According to EU legislation the body of the shark now has to be retained on board. But a mysterious loophole in the regulation still allows fins to be cut off on board, and operators to land fins in one port and carcasses in another, leading to a problem for inspectors charged with matching the number of fins aboard with the number of landed shark carcasses – controlling in other words, whether shark bodies have been illegally dumped or not. The EU decreed that the weight of shark fins on board could be at most

5 per cent of the weight of cleaned shark bodies – a decision roundly criticised by scientists and environmentalists, who claimed that 2 per cent would be a more appropriate figure.

In 2006, Rosa Miguélez Ramos compiled a parliamentary report which stated that the weight of fins could not exceed 6.5 per cent of cleaned body weight of some shark species. This prompted Fraga Estévez to chime in and say that "bad legislation" should be updated and that EU fishing fleets should be allowed to land a larger proportion of fins in relation to the number of shark bodies, in particular with regard to blue sharks (*Prionace glauca*), the commonest catch for Spanish shark-fishing boats, and red-listed as "Near Threatened" by the International Union of the Conservation of Nature, IUCN.

The non-governmental organisation Shark Alliance that works to protect the world's shark species – one third of which is on the IUCN Red List with another 20 per cent imminent – was hoping for tougher rather than looser EU regulations. An adjustment to the percentage from 5 per cent to 6.5 per cent would mean that three sharks could be "de-finned" for every shark landed, it was claimed.

"The EU finning regulation is already one of the world's weakest," noted Sonja Fordham, director of Shark Conservation for the Ocean Conservancy organisation.

It was no surprise that it was two Spanish members who took interest in the shark finning debate though. Spain is by far the biggest shark catcher and trader in Europe. The fins are sold to Asia for shark fin soup and can fetch more than €100 per kilo, with the rest of the shark of little monetary value. In 2003 Spain had the fourth largest shark catch in the world and was also the next largest exporter of shark products.

Another debate, this time about the dumping of unwanted catch, also arouses my interest while I surf the web for Committee on Fisheries documentation. On January 28, 2005, committee chair Philippe Morillon asks the European Commission about an international plan of action for elimination of discards and reduction of by-catch. According to the UN Food and Agriculture Organisation (FAO), 7.3 million tonnes of fish and other marine life not targeted as primary catch are discarded every year.

"In the last ten years, 8 million euros has gone from the annual EU budget to more than 400 research projects aimed at increasing selectivity of catches. It is now time to focus on implementation of the necessary steps to reduce by-catch," Morillon says.

Scottish MEP Catherine Stihler fills in:

"Seven point three million tonnes of fish discarded! That's a really shocking figure! It's not much lower than the total amount of fish landed by all 15 member states in 2002 to 2003."

The Swede, Carl Schlyter, also chimes in:

"If we can't find a solution to this problem, it will shortly be as empty in our seas as it is in this room right now."

But there turns out to be at least one more committee member present during this debate, and it is the Portuguese socialist Paolo Casaca who now makes an impassioned speech that ought to have a far wider audience:

"I believe that so-called by-catch more than anything is a problem of our civilisation. I find it utterly barbaric that millions of tonnes of wild animals are harvested for no reason whatsoever. This is killing for killing's sake, killing caused by bureaucratic rules and an industry mindset that makes it cheaper to kill more and use less, based on the idea that nature has no value at all. The point is that we must respect nature! We cannot have sustainable fisheries without respect for nature. Discards is perhaps the strongest manifestation of the negative side of our common fisheries policy. Besides, we need only to look at our neighbours in Norway and Iceland, where such methods are forbidden. I am horrified we are not following their example."

End of meeting.

(But not end of story. The issue of a possible discard ban will be discussed in Brussels for many years to come. According to the EU Commission's proposal for a new Common Fisheries Basic regulation in 2011 it could possibly be made reality sometime 2014–2015: if member states and parliament agree.)

I am about to meet one of the most powerful people in the field of fisheries policy in Europe. He is a Swede, his name is Jörgen Holmquist, and since 2002 he is Director-General of the European Commission's fishing directorate, DG Fish. He is in other words head of the EU "fisheries administration", a non-political authority within the European Commission.

Holmquist immediately agreed to meet me and help me sort out some of my main questions – in fact the very starting point of this whole book. Why was Sweden, in 2002, not allowed to unilaterally ban cod fishing within its own 12-nautical-mile territorial limit? How could EU legislation trump national sovereignty to the extent that, as Green Party MEP Carl Schlyter put it, there existed a "duty to exploit" our own waters? And, not least, how could the former EU Commissioner for Fisheries, Franz Fischler, who preceded Joe Borg, tell the Swedish government at their first meeting that the EU had nothing against a unilateral Swedish cod fishing ban, and just one month later all of a sudden veto it? Was there really any substance to the vague words about the "Treaty" and the exclusive competence of the CFP? Was it really true that the EU has a monopoly on decisions regarding how much fish should be taken, even when a member state wants to take *less* than agreed, within its own territorial waters?

One source has it that Sweden's decision did not become a problem in Brussels until the Ministry for Agriculture sent a note to the EU Commission requesting formal approval for the unilateral cod fishing ban. On a previous occasion in the 1990s the Swedish Board of Fisheries had banned salmon fishing in the Baltic Sea for a period and it had worked splendidly without the EU being asked, and no EU boss had called up to complain either – according to the then Board of Fisheries Director-General Per Wramner. Regarding the cod ban, the Fisheries Commissioner Franz Fischler actually had given the green light to the project orally but, according to my source's theory, the official Swedish request was deliberately phrased in such a way that the answer couldn't be other than no. The EU has a Common Fisheries Policy (CFP) which, along with the Common Agricultural Policy (CAP), is one of the major areas where the union really has exclusive competence, and a unilateral moratorium would have been a fatal step in the opposite direction of what the European Union is striving for, that is, a union with many common policies, a common constitution and a common currency.

DG Fish (its name will change later to DG Mare), is located in a dull, modern office building in a sparsely populated section of one of Brussels' many bureaucracy districts. On a rise at the end of the same street, Rue Joseph II, is the large Council of Ministers building, shaped like a peculiar screw nut with four wings. The European Parliament's great glass buildings are a fifteen minute walk away.

I am surprised to be met by Lars Gråberg, a Swedish official who will sit in on my meeting with Jörgen Holmquist. Gråberg is a veteran of DG Fish and has the title "principal administrator". He is middle-aged, wears a grey suit and is pleasant and relaxed. He tells me he works with quotas and prepares negotiations among other things, and accompanies me into Jörgen Holmquist's huge office on the fifth floor. It is a spartan, very masculine office with black leather armchairs, a large conference table and many marine-themed works of art. An odd picture of a fish skeleton and a rainbow-coloured boat grasps my attention.

Holmquist is a slim man with glasses, a slightly furrowed face and an intelligent grey gaze. He has the talent of quickly tuning into the flow of a conversation and speaks in such a personal tone that I get the impression I am being given confidences. That "important" Brussels feel that I sensed in the parliament, is very tangible in this office, with the "principal administrator" and his international boss who has just taken off his jacket.

"Let me try to explain," sighs Holmquist, leaning back in his chair when I ask him why Sweden was not allowed to try to save its own cod.

"The rules would have allowed Sweden to stop its fishermen in Swedish territorial waters. But it would not have been able to stop Danish, German or even Swedish fishermen outside the 12-mile limit. A ban would have only reshuffled the catch. It wouldn't have had any effect."

He goes on by telling me that, like their European colleagues, Swedish fishermen have the right to exploit their allotted quotas. This was further emphasised by the cod recovery plan already in place in the Baltic, based on the scientific advice of ICES, which Sweden had agreed to follow.

In those circumstances, it would not have been appropriate for Sweden to go its own way. But would a Swedish unilateral moratorium really have been *illegal* according to the EU treaty? "Yeah, well ..." Jörgen

Holmquist goes fuzzy, just like everyone else does when I raise the issue.

"It is a complicated legal discussion. But the decision belonged on the EU level."

"But would Sweden have been able to sneak in a cod ban without saying anything?"

Holmquist shakes his head.

"No. These things have to be in accordance with EU law. It wouldn't have worked. If I remember rightly, we had already begun a legal analysis after the fishing sector had complained. And then we saw the government note ..."

I ask again if the note had been phrased to elicit a no; to get the Social Democratic government out of the corner it was put in by the Green Party (Sweden's Prime Minister at the time, Social Democrat Göran Persson, later confessed he had been irritated by the Greens after the election of 2002 when they were throwing their weight around, even negotiating with the Conservative parties). Jörgen Holmquist looks puzzled, but Lars Gråberg nods in agreement.

"I kind of understood they wanted a no," he says. "It was one of the Green Party's heartfelt issues and it gave the government a dilemma. Let me put it this way: I don't think they had anything against the Commission saying no. We were mainly worried about departing from a common policy that involved duties and rights for member states. We saw it as discriminating against Swedish fishermen."

I glance again at Jörgen Holmquist, who attempts a cautious summary.

"The Green Party was very committed. The Social Democrats were lukewarm. Some Social Democrats did not want a ban, but I don't think that held true for the party as a whole."

That no one in the EU reacted in the 1990s when Sweden temporarily banned salmon fishing he puts down to "a little demonstration of EU flexibility". He won't comment on why Franz Fischler initially agreed to the cod ban. The version I heard is that Fischler was unprepared and hasty, and Holmquist doesn't contradict.

We switch the subject to the lobbyists that I assume file through this elegant office on a daily basis.

"It is strange the way that people think so badly of lobbyists," Holmquist says, "and still say we don't listen enough. Lobbying isn't forbidden. On the contrary, the Commission *should* be listening. We get visits from Scottish cod fishermen, British fishermen, environmental groups, the processing industry. We listen to all the interests. If someone wants to talk to us we always try to make time. Of course, we can't let ourselves be overly influenced by narrow sectors, but I don't think we are. Few industries are as critical of the Commission as the fishing industry."

He points out again that "the environmental movement isn't as mad at us as the industry", and there is something in what he says. It was the Commission, through DG Fish, that produced the famous Green Paper that brought the fishery problem to the attention of the Swedish Green Party among others in 2001. Its openness and self-criticism were impressive, and the environmental groups all agreed. But in European fisheries industry circles it unleashed a virtual storm of aggression; the Portuguese were angry, the French were angry, the Italians were very angry, and the Spaniards, well, they were livid!

The spring of 2002 in fact staged a real docu-soap opera in Brussels because of the Green Paper and the upcoming reform of the CFP. In a series of articles, the Swedish newspaper *Dagens Forskning,* reported on the weird machinations and plotting and planning in the Brussels corridors that led to Jörgen Holmquist's predecessor, Denmark's Steffen Smidt, quitting his office on 24 hours' notice with no explanation. An informal group calling itself *Les Amis de la Pêche* ("Friends of Fishing"), comprising France, Italy, Portugal, Greece, Ireland and Spain did every-thing it could to obstruct the reform that would to be presented in April of 2002. The problem from their point of view was that the new policy would end support for new ship construction, and cut the size of the total EU fleet by 8.5 per cent.

According to *Dagens Forskning,* the means used by *Les Amis de la Pêche* to kill the reform were unusually blunt, especially from Spain. The Spanish Transport Commissioner Loyola de Palacio was reported as having made a secret call to fisheries commissioner Franz Fischler, and Spanish Prime Minister José Maria Aznar was also said to have secretly contacted the then Commission chair Romano Prodi to try to make a deal, perhaps by backing the Italian city of Parma's candidature to host the new EU foodstuffs authority. Meanwhile, Spain's minister of fishery, Miguel Arias Cañete let slip in a television interview, and not at all secretly, that "we have mandated our respective commissioners to try to stop the reform." He was reassuring Spanish fishermen; and the "we" he was referring to was *Les Amis de la Pêche*.

Later he withdrew his statement, claiming there had been "a misunderstanding".

Probably the actions by the Friends of Fishing did have some impact. At least the presentation of the reform was postponed by a month. Subsidies for new boats were allowed to continue until the end of 2004. And for some reason Director-General Steffen Schmidt was fired. But why? According to some the Spanish made him a scapegoat, accusing him of being behind what they saw as the overly radical reform. Others say that he was on the contrary too weak, and that Fisheries Commissioner Franz Fischler needed a stronger person at the head of the Fisheries Directorate to help him push through the reform.

Both theories might be true. But the choice of the bespectacled Swede sitting across from me as a successor to Schmidt indicates that the latter theory might be more correct. In any case, Holmquist's appointment did not mollify the Spaniards. "We are surprised that the European Commission again appoints a Director-General from Northern Europe. We are concerned about the imminent negotiations for the fisheries reform," said Javier Garat Perez, Secretary-General of the Spanish fisheries organisation FEOPE, speaking to the Swedish journal *Riksdag och departement* (02-08-2002).

Holmquist himself pulls down the shutters when the furore over his predecessor is brought up.

"Spain put pressure on the Commission, but I think the situation was more complex than that. It is an extremely sensitive issue for the Commission to act when a member state exerts pressure."

We finish our conversation by looking ahead. What's going to happen? I wonder. Are there any radical solutions up the Commission's sleeve, or will the new 2007-2013 programme period be more of the same? Will fishing be regulated via quotas, is that the best way? Should discards be banned? What's happening with the ecosystem approach?

Holmquist answers knowledgeably, with the usual reservations: quotas have advantages and disadvantages, and "relative stability" is important, meaning that each country should be getting the same percentage of the shared stocks as before (a share based on a percentage of catches of shared stocks in the 1970s). Discards (or dumping) is not good, but we should beware of creating a market for juvenile fish, which might result if fishermen are allowed to land small fish. The ecosystem approach – meaning taking the entire ecosystem into consideration when estimating how much a particular fish stock can be fished, to not just count cod but also herring and zooplankton and phytoplankton when deciding on sustainable fishing quotas – well, that is an issue that prompts the shirt-sleeved Director-General to smile and crack a real fisheries-bureaucrat joke:

"It's extremely complicated to analyse stocks using the ecosystem approach. Our worry is that scientists who do stock assessments are now a dwindling stock!"

The principal administrator and I laugh politely. It's getting late and I thank the two for a very interesting talk. I am on my way to my next stop in Brussels: the European Parliament hearing of the designated new Fisheries Commissioner, Joe Borg. Completely anonymous in the world of fishing – a former university law lecturer from Malta, the smallest of the EU member states – this man will soon become the most important figure in fisheries policies in the industrialised world, provided he comes through the cross-examination. Jörgen Holmquist, meanwhile, would remain the second most important figure until the beginning of 2007 when he became chief of the EU Internal Market and Services Directorate General. In 2011 he was to return to the fisheries sector as the new chairman of the board of the EU Fisheries Control Agency Board, situated in Vigo, Spain.

On TV monitors in the large press section with hundreds of temporary working spaces for journalists, I see the round face of Italy's Rocco Buttiglione wherever I look. I also catch a glimpse of Sweden's MEP Maria Carlshamre, who seems to be asking a critical question to the Italian candidate of being Commissioner of Justice but I don't have time to sit and listen. After the interrogation of Buttiglione it is Joe Borg's turn, and I leave my coat on a working place in the press room, and hurry off to find Room 1A2 in the Paul-Henri Spaak Building, where Borg is to be quizzed.

To my astonishment the hall outside the chamber is packed with people. I recognise a few faces, among them Director-General Jörgen Holmquist, standing alone, casual looking and waiting to take his seat inside. Otherwise the crowd mostly consists of men and women in suits who are conversing loudly. There is excitement in the air and a crush to get in when the parliament hostesses open the doors. My fear that I won't be able to find a seat quickly evaporates: the chamber is large and there are many places reserved for non-MEPs. "Presse", "Invités", "Corps diplomatique", "Représentations permanentes" and "Conseil de l'Union Européenne" is written on white cards with black lettering and I find a seat in the press section, beside a couple of colleagues.

On a TV monitor against a blue background, a text announces the three-hour hearing with the "Commissioner-Designate", that is, candidate Borg. Two cameras are pointed at the table in front of the blue backdrop where he will shortly be seated.

The grey-haired, balding Borg vaguely resembles former Soviet leader Mikhail Gorbachev as he takes his seat and begins his initial statement in his native Maltese.

The hearing is to last three hours and the procedure has been meticulously planned. According to the agenda I have in my hand, Borg has already responded in writing to questions about himself and his view of European fisheries policy. The time is strictly allotted to the various political groups. I note that the large Conservative group EEP, whose deputy chair Carmen Fraga Estévez I have already seen in the Fisheries Committee, has been allotted 44 minutes, the Social Democratic S&D group 32 minutes, and the Greens/EFA 8 minutes. I spot the corduroy back of Swedish Green member Carl Schlyter in the front row of the chamber and wonder if he will get an opportunity to ask a question.

The hearing gets underway and the questions sweep quickly through subjects like discards, fisheries agreements with third countries, improving control over illegal fishing, the impending Regional Advisory Councils, and subsidies to the fishing fleet in the 2007-2013 period. It's as though everything and nothing is being said at the same time. I listen to the diligent Swedish interpreter who manages to sift through Joe Borg's often evasive answers to produce: "Striking a fair balance ..., setting up objectives ..., taking into consideration ..., abiding rules ..., in dialogue with ..., weighing the interests ..., finding financing...".

Oddly enough, Borg's extremely balanced answers manage to irritate almost all the members, and the initial congratulations and well-wishing of the parliamentarians quickly turn into challenging questions. Does the Commissioner-Designate intend to compensate the fishing fleet for rising fuel prices? How does the Commissioner-Designate see the future of special support for fisheries in the EU's ultra peripheral regions overseas? How does he view the fisheries agreement with Norway? What are his thoughts on the recovery plan for southern hake? What is his view on the "social dimension" for fishing communities? How will he ensure that the biological data underpinning stock assessments are new and accurate? And will he work to restart the EU negotiations with Morocco on a new fisheries agreement?

Joe Borg dodges the bullets skilfully, although no doubt disappointing many MEPs. He does *not* believe in reintroducing financial support for the construction of new vessels in the coming period. He will not "wag a finger" at the scientists, who are doing their job. He thinks that EU fisheries agreements with third countries must include more compensation and support to local fisheries sectors in the developing countries concerned.

"There has to be an end to 'pay, fish and go,'" he says. "It's wrong. And European vessel owners will have to pay more for the access to third countries' waters."

In his response to Rosa Miguélez Ramos, the Spanish socialist delegate who had raised the Morocco issue, he is quite clear:

"I am not sure that Morocco *wants* to negotiate. That makes it hard to get a new protocol."

After additional questions on Borg's view of fisheries in Poland, Galicia and in the Azores, time has finally come for Sweden's Carl Schlyter to put a question. He takes off his earphones and speaks in Swedish, raising the issue of the aborted cod ban of 2002. I prick up my ears.

"If a member state finds the EU's conservation measures insufficient, shouldn't they be free to take unilateral conservation efforts in their own waters?" he asks.

Joe Borg's answer echoes Holmquist's earlier that day: the Commission was against Sweden's unilateral cod fishing ban because decisions of that nature lay within the exclusive competence of the common fisheries policy. Besides, a recovery plan already existed and there was no cause for unilateral measures.

Schlyter asks for the floor again and puts his question even more succinctly:

"Are you against a country abstaining from its quota? If the Commission is against a member state voluntarily abstaining, I have to interpret that as an obligation, a *duty* to exploit fish. Is that correct?

I look intently at the Commissioner-Designate as he listens to the interpreter, predictably pulling a face when he hears the words "duty to exploit". When he finally speaks, he is reiterating Holmquist's explanation.

"If a country relinquishes its quota, it can be shared between other countries. We have to look at this and think about it but I believe it could be hard to find a solution where individual countries can abstain."

By this stage, the Green group's minutes are over and none of the subsequent questions concern Sweden. But I stay and while I listen to endless "taking into consideration ..., all stakeholders ..., objectives ..., rules ..., dialogue ..., financial sources ...". I think of Bo Hansson, the fisherman from Smögen on the Swedish west coast, and all his coastal fishing studies and reports. How much of his thoughts penetrate through to these forums? Or how much of scientist Henrik Svedäng's full-time work reaches this political level?

On my way out, I buttonhole Sweden's only directly elected representative in the Fisheries Committee and ask him if he's free for lunch the next day.

He is.

We meet by the sculpture outside the Hémicycle, the large main plenary of the European Parliament in Brussels, and take the lift up the 12th floor and the airy MEPs' canteen, with its high ceiling and wide view of Brussels. In an EU that often distributes grants to those not so much in need, it's only logical that the well-paid MEPs also enjoy a heavily subsidised lunch. It costs only €4.20, half of what you pay in a normal restaurant in the city. And there is a large choice of dishes, plus salad and bread.

"The prices went up a while ago and the MEPs complained," Schlyter confides with a grin.

We find seats and he tells me he chose to be substitute member in the Committee on Fisheries because he is married to a Greek and has seen the effects of overfishing in the Mediterranean. But he is a full member of three other committees and this is not his main job.

"To be honest, there's little interest in the Committee on Fisheries in parliament; most people want to be on sexier committees. It's more fun to get involved in forests than in coral reefs. Fish can't scream they have no soul; no one cares about fish. It would have been different if fish were cute like dolphins. It would be easier to get the public involved, and it would score more political points."

He says he is still in a learning curve regarding fishing issues. He gets briefings from the Green group's fisheries adviser Michael Earle, much as Joe Borg was briefed yesterday by his experts. Schlyter says he thinks Borg seems all right.

"The Commission is the radical force in European fisheries politics. Parliament and the Council of Ministers are the resisters."

I wonder how much MEPs are influenced by fishing industry lobbyists and Schlyter says there are close ties between industry and MEPs.

"Industry lobbyists have always projected the attitude that of course you'll have time for them," he says. "It's not the same for the others, the non-profit, non-governmental organisations."

He says he tries to fit in a meeting with a non-governmental organisation, a NGO, or an independent scientist, for every company lobbyist he meets, to balance impressions.

"Of course the lobbyists influence us! It would be naive to deny it. Otherwise they wouldn't be trying!"

It turns out Schlyter doesn't have much to tell me about fisheries that I don't already know. He confirms that the Spanish dominate the committee and that the EU fisheries agreements with the Third World are "morally reprehensible".

"Why do they still exist? Because Spanish government members don't want fishermen coming round throwing fish at their homes."

He regrets that fisheries agreements are routinely sent to Parliament *after* being negotiated, signed and already in operation. Strong forces are intent on avoiding a discussion, he believes.

"Remember that a large percentage of the tuna fished in the world is taken by the EU. It's a lot of money."

After finishing eating we take our trays, walk through the restaurant and I remark on the high, greenhouse-like round window arches, trademark of the European Parliament from the outside. We stop for a while, gazing out at the panoramic mile-wide view of Brussels and suburbs.

"They want this to feel like a ship's bridge," Schlyter says. He nods out the window and exclaims ironically:

"You know: from up here we are steering Europe!"

AQUACULTURE – THE BEST SOLUTION?

Öksfjord, northern Norway, September 2005.

The water is completely flat, like a mirror. At the fjord entrance a small fishing boat is approaching. Wispy white clouds hang over the mountains and all is reflected in the sea: the blue sky, the yellow-green rock faces, the seagulls, the teeny red boat. As far as we can see, the landscape is completely untouched – and numbingly beautiful.

We are in Öksfjord, not far from the North Cape and the Barents Sea, home to the world's largest remaining cod population. But even here the cod stocks aren't safe. For more than a decade the number of fish that die as a result of fishing has been far higher than the precautionary principle allows. Catches have gone down from 1 million tonnes in the late 1960s and early 1970s, to less than half a million. At the same time the capacity of Norway's fishing fleet has increased by 70 per cent.

Besides over-intensive legal fishing by Norwegian and Russian trawlers, the Barents Sea has also suffered from widespread illegal fishing, estimated by the International Council for Exploration of the Sea (ICES) at a massive 166,000 tonnes in 2006 (double the amount of cod caught legally in the Baltic Sea). There is, to be sure, always the risk that the Barents Sea cod may suffer the same fate as that on Canada's Grand Banks. Responding to these fears, ICES put the brakes on for 2007, recommending a one-third cut in Barents Sea cod catches. And recommendations for coastal cod catches have been the same for years: zero catch. The local stocks living in the fjords are, according to ICES, smaller and more vulnerable than ever before. (In 2011 the coastal cod populations were in its lowest ever recorded status in spite of a long-term recovery plan, while the Arctic cod has recovered tremendously, almost doubled in population in just six years, mainly thanks to stricter controls on illegal fishing.)

But the Norwegians have a special ace up their sleeves in the struggle to preserve hard-pressed wild fish stocks, and for more than three decades they have been fine-tuning their methods. If the solution to the world fish crisis is aquaculture, well, then the solution is to be found in Norway. The Norwegians began small-scale fish farming in the late 1960s after salmon and salmon trout numbers plummeted due to the expansion of hydropower in the rivers, and farmed salmon now accounts for almost half of Norway's fish production.

The plan now is to repeat the salmon success story by farming cod, and I am in Norway to visit one of the new cod farms – an industry grappling numerous teething problems. First, however, we have stopped in Öksfjord to look at a salmon farm – an industry that has been going for so long that one would have thought it ought to have overcome its problems by now. With more than 800 salmon farmers, Norway is the world's largest producer of cultivated salmon. Chile is second, producing about half of Norway's output, while the other leaders – the UK, Canada and the Faroe Islands – lag well behind. Norway's long coast and innumerable protected fjords are its key to success. Aquaculture in the open sea is much more difficult, because storms often lead to equipment damage and escapes, while the fjords, which offer protection from high waves, provide a much more beneficial environment. Also because the many rivers that run into the fjords provide a healthy flow of water through the big cages where the fish is kept, diluting the high concentrations of waste – excrement, residues of feed and medicines – that the farms generate. In more densely populated areas this waste would cause major environmental problems; a salmon farm with 1,000 tonnes of fish emits as many nutrients in the water as 10,000 people. In other words, Norway's fish farms contribute to water eutrophication as much as the entire Norwegian population.

I am sitting in a small humming outboard motor boat with two men in padded boiler suits, heading for six circular pens in the middle of the glassy fjord. The men work for the Volden Group, a small Norwegian salmon producer. The company has a concession to farm salmon here, and at seven other locations, and has total annual production capacity of 10,000 tonnes – a small fraction of Norway's national salmon output of 500,000 tonnes. We leave behind two tiny crests of waves that break the mirror surface, and after a few minutes I start to see the green nets more clearly, and the salmon that are jumping inside.

"The salmon are infested with lice. That's why they jump so much," one of the men explains, breaking the spell that for a moment had me almost buying the company's advertising blurb, the one about "10,000 years in the wild, Arctic landscape":

In these remote regions, the salmon and trout develop a quality that is unique. It is here, in the midst of such wild and majestic scenery, with depthless, icy fjords and a light that quickens the blood of man and beast, that the sea forges its own silver – Blue Silver™. For generations, the name of the Volden Group has been synonymous with the outstanding quality of its salmon and trout – firm, red flesh, clad in blue and silver.

Accept our invitation to a taste experience that is the culmination of 10,000 years of history and tradition in the wild Arctic. We give you only the very best and we take pride in presenting you with a unique taste sensation – a taste of Blue Silver™.

We have anti-louse medicine with us in the boat, the idea being to put it in the water to make the fish slimy, and prevent the lice from getting a grip. The small bloodsuckers like to settle in the gills, which irritates the salmon enormously. The fish jump and flip incessantly in the cages, every time landing sideways with a splash on the water's surface. Jumping and letting the gills slam the water is as far as I understand the fish's desperate attempt to scratch.

The salmon louse (*Lepeophtheirus salmonis*) is one of countless parasites and ailments present in Norwegian salmon farms. The lice occur naturally at sea but thrive of course where there are large concentrations of fish. These Volden salmon pens, with a circumference of 90 metres and a depth of 10 metres, hold 25 kilos of salmon per cubic metre – and are a virtual paradise for the tiny brown bloodsuckers.

But the main concern about the salmon louse from an environmental point of view, is not that it is plagues the poor farmed salmon, but rather that it also infects wild stocks. This can occur when cultivated salmon escapes and interacts with wild fish, or when the louse itself sets adrift to find new host animals. A University of Auckland study has shown that salmon lice can swim up to fifty kilometres at sea to find a new host. And another study published in the *Proceedings of the National Academy of Sciences* of the U.S. found that lice are lethal to the wild salmon smolts leaving their native rivers; just a couple of lice suffice to kill a juvenile salmon. According to the study, in Canada between 9 and 95 per cent of wild smolts have been infected – an alarming figure for wild stocks already under duress from human activity. This research is one of the most extensive in its field, and in line with the findings of similar studies in Norway in 1997, when the Norwegian Institute of Marine Research and the Institute of Zoology at Bergen University found that almost *all* outward migrating wild salmon smolts in Vestlandet – an area with more than a hundred million cultivated salmon – died from lice infestation. The mortality rate at Sognefjord was nearly a hundred per cent and at Nordfjord around sixty per cent. Remedial measures have since been taken at Sognefjord to radically reduce the death rate, but the problem is far from over. The danger is that aquaculture, which many people hoped would take pressure off the wild salmon, could be the death blow for the species instead.

A grey decked iron boat stands moored alongside the pens out in the fjord. Inside it are all sorts of equipment to automatically monitor and feed the fish. Once a minute a batch of brown pellets, looking like dog food, spurts out of a pipe above each pen. But instead of rising to the surface to try and catch them the fish keep swimming round and round at a steady pace; probably because the water is already so thick with feedstuff that the fish can eat at any time; only by opening their mouths. This is how farmed salmon live, for two years, until they have grown to four kilos and are big enough to be slaughtered. Farmed fish are protected under the Norwegian Animal Welfare Act, which means that the slaughter has to be conducted under certain conditions in a special salmon slaughterhouse, to which the fish is transported in a special tanker. Before being slaughtered the fish has to be anaesthetised.

The brown pellets being sprinkled on the water surface are composed mainly of fishmeal and fish oil, dosed with liberal quantities of the synthetic pigments astaxantin and cantaxantin, to make the salmon pink. Wild salmon absorb astaxantin naturally by eating small crustaceans, but farmed salmon must be fed the substance artificially or else their flesh becomes unattractively pale, and in addition they end up with vitamin deficiencies.

The main problem however, when it comes to the feed of farmed salmon, is not the artificial colouring (even though it is questionable that we allow fish to eat a substance that is not allowed directly in human food) – but something completely different: the huge amount of caught wild fish used as farmed salmon feed. Experts differ on the figures: some suggest the salmon eats two to three times more wild fish than the output of farmed fish, while others put the figure much higher. According to Norway's salmon farming industry, the ratio is just over one kilo of wild fish for every kilo of farmed fish, though this is an over-optimistic number that the industry itself played down on Swedish television (*Uppdrag Granskning* – "The Pink Gold" – February 25, 2009). It turns out the industry relates the *dry* weight fish feed to the *wet* weight of farmed fish in an attempt to improve figures – but even if it were not the case, the relentless growth in fish farming is becoming a serious problem. Based on current trends, the FAO (United Nations Food and Agriculture Organisation) forecasts that soon all "feed fish" (small pelagic used to feed animals) caught in the world will be required as food to cultivate predatory fish species, like salmon and cod. A more sensible solution according to the FAO would be if Westerners could change food habits and eat more plant-eating farmed fish such as carp and tilapia (provided they are sustainably cultivated). By the way, carp is the most popular species for cultivation in China; the world's number one producer of farmed fish.

Another, more energy-efficient solution would be of course if we could consider skipping a step in the food chain and eat the feed fish directly ourselves. In Scandinavia, these species include herring, sprat, sand eel and capelin, and at the moment we consume their protein and healthy omega-3 fatty acids only via a, to say the least, circuitous route that runs from Swedish industrial trawlers in the Baltic, via fishmeal factories in Denmark to salmon farms in Norway – and sometimes also passing through Asia for filleting and packaging before being transported back to Sweden and the end consumer in the food stores.

But how healthy are the animals really in the Norwegian salmon cages, after more than thirty years of experience in aquaculture, after so many years of trial and error? On the positive side, the once alarmingly high use of antibiotics has in fact now dropped to a small fraction of what it was during the late 1980s and early 1990s – thanks to new vaccines that protect the fish against many diseases. The problem is that new diseases are emerging all the time. Recently the Norwegian National Veterinary Institute reported a massive rise in the mysterious Congenital Myasthenic Syndrome (CMS), which causes heart failure in adult salmon. Sixty-nine cases have been recorded along the Norwegian coast but the authorities suspect the true incidence to be much higher. Other diseases reported by the Institute of Marine Research are the infectious salmon virus ILA at 16 fish farms, pancreatic disease at 44 farms, Infectious Pancreatic Necrosis (IPN) at 172 farms and the heart and skeletal disorder HSMB at 54 locations. In addition to these there are also the common bacterial diseases such as winter ulcers and boletus, parasites such as salmon lice and worms, injuries and the negative side effects of vaccination such as peritonitis and deformities.

The majority of these ailments can of course also infect wild salmon, and they do. Only one week before my Norwegian visit, two separate large fish farms were wrecked by accident, resulting in thousands of salmon escaping. According to the salmon industry's statistics the escape rate sounds innocently low at only 0.3 per cent. But given that Norway produces 300 – 400 million farmed salmon a year, it actually adds up to almost one million escapes – almost all of them carrying nasty germs and viruses, all too keen on jumping on to new wild victims.

Another problem with the escapes is that the farmed salmon interbreed with their wild cousins, weakening the gene pool. The farmed salmon are descended from a narrow lineage, specially designed to grow fast, and in the wild the risk of inbreeding is permanent. And with their shorter, faster growing bodies and smaller heads, cultivated salmon are a lot less smart, and less able to survive in the wild than wild fish. And they are not particularly socially skilled either: gangs of farmed salmon have been observed behaving like hooligans; disturbing their wild relatives at their spawning grounds.

Today nobody knows how many Norwegian salmon in the wild actually are hybrids between wild and farmed salmon, but the Institute of Marine Research is investigating the matter. Already it is quite clear though, that the genetic differences between the thickset, farmed salmon flailing helplessly in their cages to try to escape the lice, and the silvery, slender wild salmon in nearby rivers, will soon be very small.

After we have inspected the feeder boat and through an underwater camera watched the tireless salmon swimming round the pens, something happens. As we approach the shore in the boat, out of the corner of my eye I see what I first believe to be a big fish breaking the flat water surface in the fjord. And then another! Then I clearly see two dark, shiny, rounded backs with small dolphin-like fins rise repeatedly to the surface, side by side, spout and dive again. They are porpoises, a species long been on the IUCN Red List of threatened animals classified as "vulnerable". This species is almost extinct in the Baltic Sea and severely depleted along the Swedish west coast. Since the dawn of time this little mammal lived on the generous amounts of fat herring in Scandinavia's icy waters, until it started declining in the 1970s, most likely due to the overfishing of herring stocks and possibly also to organic pollutants that might have affected its reproductive capacities. Today the herring is back and the porpoise is protected, but the problems remain since it continues to get caught by accident in trawl nets, in drift nets, or in nets on the sea floor. The number of porpoises that drown each year entangled in fishing gear is high, and most of the casualties are young porpoises aged three years or younger. Since the species reaches maturity as late as at 3-4 years and females give birth to a single calf only once every two years, every lost juvenile is a serious blow to the threatened population.

Probably we saw a mother and calf, attracted into the fjord by a shoal of herring my companions tell me. I stand watching the smooth, empty surface as we close in on land, but the porpoises never reappear. I wonder if the glimpse of these small, rounded dark backs was the first and last time I will ever see a porpoise in the wild – and, as so many times before,

I marvel that one single glimpse of a wild animal can arouse such joy in a human being.

I take the famous *Hurtigruten* ferry to Havöysund, a fishing village of 1,300 people, one stop away from the North Cape. Here the air is teeming with seagulls, the sea is a dark blue and the red and white fishing boats are much smaller than I expected, considering these are the richest cod fishing grounds in the world. Gerhard Olsen, chairman of the Havöysund fishing organisation, invites me on board his boat that lies at the deserted quay. The wheelhouse is warm and comfortable but small, with space enough for just two people; very small when you consider that just a few minutes away from the safety of this harbour you will have nothing but open sea and ice between yourself and the North Pole.

The instruments on board are ultra-modern, though, and Gerhard shows me the trawl lines of the last few years on his computer – red lines, resembling tangled balls of yarn against the turquoise-coloured Barents sea on the screen. The fishermen of today have great use of modern technology, Gerhard says. Coordinates of every trip can be recorded, which enables them to draw from experience, and return with greater certainty to where the cod schools were in the same period the previous year.

Gerhard offers me a cup of instant coffee, and with his traditional knitted Norwegian sweater, beard and blue eyes he sure looks very amiable and jolly. But he says that he worries over some things concerning the future of the cod fishery.

"Oil tankers," he says. "Supertankers of 100,000 tonnes pass through here. Just imagine what would happen if there was a major oil spill in the cod breeding grounds, it'd be catastrophic."

Then he tells me about illegal fishing, singling out Russian ships as the main culprits. The Norwegian coastguard has a nigh on impossible job monitoring the vast ocean containing fish resources worth billions of Norwegian kroner every year. But the Russians are not the only offenders; Portuguese and Spanish boats fish here illegally too. Things have improved slightly in recent years, Gerhard acknowledges, thanks to rules forcing larger vessels to be fitted with VMS, satellite transmitters.

"A lot of fishermen were against them at first, but we think they're good now. There's so much money at stake. A seine boat can catch a hundred tonnes of fish in a single cast."

Not far from the dockside is the Havöysund fishermen's centre, where Gerhard offers me more coffee and some brochures from the Norwegian Institute of Marine Research. The walls are lined with black and white photographs of men in sou'westers holding up massive specimens of cod, and Gerhard explains how fishing dominates the entire community; the fishermen's wives are organised in a group of their own, and the Tobö fish processing plant in the centre of the village employs eighty people, mainly women and young people. Four million kilos of frozen white fish and fish fingers a year are exported to the US and Europe from this processing plant alone. Fisheries are the life and blood of the community, and Gerhard seems genuine in his hope that the world's largest cod stocks in the Barents sea will not be overfished, as it would be in nobody's interest, certainly not the fishermen's.

Alas, there are indications that the Arctic cod could also be in trouble. Catches are dropping despite more intensive fishing, while the average age of caught fish is also dropping; a sure sign that fishing pressure is too high. There are practically no giant cod left like those in the photographs anymore, and it is rare to catch fish even half that size. Young cod are plentiful but the stock is showing signs of adapting to increased fishing pressure by reaching sexual maturity at an earlier age – a disturbing phenomenon I first heard about on board the Swedish research ship *Ancylus* and which was also recorded on Canada's Grand Banks in the early 1990s. In the past, cod in the Barents Sea would not begin to reproduce until reaching nine years of age; now they do so from the age of six. The explanation is that among the fish in the stock, those that remain small and mature early (the runts, as it were) tend to avoid capture before they spawn – at least more frequently than those that grow fast and mature late. These runts are the ones that pass on their genes to the next generation. The Barents cod that once could live to 30 years or more and reach the length of a fully grown man, these days almost never get a chance to grow much beyond the legal catch limit of 47 centimetres. This tendency to let the smallest individuals survive to breed, and removing those with large growing potential has been described as some sort of perverted retrograde breeding programme.

And the genetic shift that this induces is counterproductive in more than one way: not only do you have to kill more individual fish for each tonne caught, but also the "birth rate" of new cod goes down since younger females lay fewer eggs than more mature females. A first-time breeder lays only around 400,000 eggs, while older females can lay up to 15 million eggs!

Gerhard tells me that Havöysund has around 100 active fishermen, ranging from sole fishermen who only fish during summer to crewmen working on the large Danish purse seine boats.

The tradition of fishing the Arctic cod (*skrei* in Norwegian) is a long and proud one, with the Norwegians having supplied *bacalao* to the Portuguese, Spanish and Brazilians and *stoccafisso* to the Italians for centuries. Norway is the third largest fish exporter worldwide, earning close to four billion euros from the industry every year. But if you scrutinise the fish export statistics, you soon discover that fish farming accounts for an ever-growing proportion of these revenues. Which is why it is not surprising that Norway, facing a potential decline in the wild cod catches – has made a determined effort to invest in cod farming in recent years. And this is what has brought me to Havöysund.

I bid farewell to Gerhard and head off to try and find Myrfjorden, a small fjord a few kilometres away from the harbour. The landscape is treeless at this latitude; not even the tiniest bush is to be seen. Here in these dark, chilly waters, former shoe retailer Aldor Johansen keeps five pens of farmed cod. I have seen numerous salmon farms on the way, and when I approach Myrfjorden the difference is visible from a distance: while the salmon flail agitatedly at the surface, these pens are calm. As we approach the pen in Aldor's boat, I notice yet another difference: while salmon swim round and round at high speed, as in a forever-spinning centrifuge, the demersal cod are less driven by instinct and are more solution-orientated and exploratory by nature. According to the Institute of Marine Research an estimated ten per cent of farmed cod manage to escape from captivity. Aldor tells me they are inexhaustible in their search for ways out of the cages.

"But we haven't had any problems with escapes here yet," Aldor assures me. "Sometimes we get wild fish in the pens instead, maybe juvenile cod or saithe looking for food. After a while they can't get out anymore, they've gained weight and can't get through the mesh."

Inside the pens, where the cod are soon ready for slaughter, the activity is completely different from that of the salmon. A slow, dignified procession of cod patrol the surface sedately; some swimming against the crowd, others taking a diagonal course before diving out of sight. Many pause to look up at us, more than one actually sticking its head out of the water and staring. Unlike the salmon, you can attract them with fish pellets. None of them jump, but a few swim at a strange angle, exposing their bellies. The mortality rate of farmed cod is 12–15 per cent, Aldor says. It is unclear why this figure is so high; cod farming is still in its infancy, and little research has been done yet. But one knows that cod are potential hosts for more than 150 parasites, and that fungal, viral and bacterial infections can do great harm to them. In the reports from the Institute of Marine Research you can see close-up photographs of common symptoms: red gills, white skin spots, damages to the skin looking like burn wounds with red and white blisters. For all its tiny size, cod farming accounted for half of all veterinary penicillin prescriptions in Norway fish farms in 2005.

Aldor throws in some feed to attract a cod, then scoops it up in a net to show it to me on deck. The fish looks large and sturdy, but has red, circular sores on its belly that could indicate illness.

Aldor doesn't take any notice, but rather wants to talk about another problem: cannibalism.

"You have to ensure the fish in a pen are roughly the same size, otherwise they eat each other. We've got better at this now and have fewer problems."

He puts the fish back, looks out to sea and readily admits the business isn't yet very profitable. Out in the ocean there are billions of beautiful, healthy, large cod available for free; all this is investment, licences and a lot of hard effort.

And unsolved problems. One is the spinal deformities that many cod suffer from, a problem that researchers have been working for years to solve, and which seems to be a question of the feed (cod fry need live food and cannot, unlike salmon, immediately switch to dry feed once they finish their yolk sacs).

Another issue is how to prevent farmed cod from becoming sexually mature in the cages – not only because sperm and eggs from farmed fish then spread uncontrolled in the wild, but mainly because spawning results in reduced weight, and the fish is then considered unmarketable. They are now experimenting with illumination of the pens during the winter, which might help delay sexual maturity, Aldor says, but – and he looks out at sea where the wild fish reproduce without interference from man: it requires a lot of work.

But clearly, the commercial success of salmon farming and the global decline in wild fish stocks still make cod cultivation potentially interesting. The Norwegians are also experimenting with other species, including haddock and wolffish.

Nevertheless, cod is undoubtedly the riskiest candidate for fish farming. If anything goes wrong here at the small Myrfjorden farm, and a serious disease were to spread to the great Arctic wild stocks nearby, it would put billions of kroner and incalculable ecological value at risk. I can see with my own eyes that infections can be spread even without the cod escaping their cages: wild cod swim just outside the pens eating the pellets that continuously fall out through the net – or possibly just showing curiosity in their captive fellow cod.

Aquaculture's potential for unpleasant surprises became extremely clear in winter 2006 when Codfarmers, the first cod farming company to float on the Oslo Stock Exchange, radically wrote down the value of its "biomass" after discovering the new and deadly cod disease caused by *Francisella sp.* in its production sites. The illness appears to be highly unpleasant; it resembles chicken pox and causes body rashes that spread into the fish's eyes, mouth and throat. Dissections also show white spots on the liver, spleen, kidneys and even heart – making the internal organs look like salami.

According to the Norwegian National Veterinary Institute in Bergen the bacteria was identified in samples from farmed cod for the first time in 2005; until then it had never been recorded in fish or mammals – though it belongs to the same family as rabbit fever (*Francisella tularensis*), which afflicts small rodents.

The devastating effect this new disease could have on wild cod would be a virtual nightmare, so it is with some surprise that I chance on an article in the Swedish National Veterinary Institute's magazine *SVA-vet* confirming a case of the disease in wild cod from the Swedish west coast, not far from Norwegian waters. The discovery followed a report from Swedish fishermen of unusual sores and organ damage in one of their catches. The Swedish Board of Fisheries examined the fish and conducted sample catches but did not find any cod that appeared to be suffering from the disease except very locally, where up to 20 per cent of cod caught were found to be suffering from skin sores. And the diagnosis was *Francisella sp.*

The news reached the media in early 2007 and Anders Hellström, head of the fish section at the Veterinary Institute, warned that the disease might spread and even threaten Baltic stocks. It remains unclear if this has actually happened because no money has been available for continued studies. Hellström laments the authorities' unwillingness to commit funding, having tried and failed to persuade the Board of Fisheries to examine the fish it catches during counting surveys and the Swedish Environmental Protection Agency to provide financial support. In fact, there is no money available in Sweden to monitor diseases on wild fish at all. And searching the Board of Fisheries' website, there is not one single hit on the word "francisella", although this new fish plague could develop into a major threat to already dwindling coastal cod populations.

Fish farming now accounts for around a third of global fish and shellfish production (though over 2/3 of it is in China). Key questions remain: is there scope for this figure to rise? Could fish farming be the solution to the world's declining wild fish populations? There are no easy answers to these questions. The main, totally overshadowing problem is of course that the equation farmed predatory fish and fish feed doesn't add up, since farmed fish requires more wild fish to eat than it produces. By contrast, farming plant-eating species still offers major potential. Tilapia, for example, is gaining popularity in the US. Farmed carp, a delicacy in Asia and in Poland, could catch on in Western Europe if we are prepared to change our eating habits.

The risk of infection and escapes from fish farms still remains a danger even after over 30 years of Norwegian experience. Unless concerted efforts are made to tackle these problems there is a risk that fish farming will not be the salvation of the wild fish, but instead threaten the health of wild fish stocks. If this happens, we consumers will have to confront the fact that our oceans no longer are home primarily to wild animals, but have rather become cultivation areas for farmed fast growing fish.

The cod with the small red sores on its belly is put back to the depths of the pen and it is time to leave Myrfjorden, first by boat, then by car. The fish farmer Aldor is despite everything rather content with his business, he tells me. Havöysund with its harbour, school, church and hotel – all rebuilt after the Germans burnt down the whole town in 1945 – needs fisheries for its future survival. And if wild stocks were to run out, farmed fish would provide a secure income.

"We'll just have to see how things turn out," Aldor says. "I'd had enough of selling shoes after 28 years anyway. This is much more fun."

THE WAY FORWARD

The EU Commission's plan for saving the eel that was presented at the Stockholm conference I attended in late 2005 was rejected almost as soon as it saw the light of day. The somewhat philosophical goal of saving 40 per cent of what could have been the amount of eels, had it not been for human interference, was kept though, but not the idea of allowing fishing only between the 1st and 15th of every month (a proposal that ignored the eel's reactive association with the lunar cycle). This idea was unanimously rejected by the industry, experts and politicians, and of course was also panned by the European Parliament. Spanish member Carmen Fraga Estévez summarised the general opinion in a debate on May 15, 2006, calling the proposal outright "absurd".

This badly prepared proposal resulted in delaying the protection of European eels another year, and the joint European eel recovery plan was not on the table until 2008. But in Sweden, the Swedish Board of Fisheries wanted to pre-empt the EU plan and address the eel's threatened extinction immediately. Its solution, announced in a press release of December 2006, appeared very radical: "The Board of Fisheries has decided to ban eel fishing in Swedish waters from 1st May 2007."

But any animal conservation euphoria that might have arisen from this apparently unequivocal ban was dashed already in the next sentence: "Exceptions to the ban will be made, however, for coastal and lake fishermen who are economically dependent on eel fisheries."

Exempted were thus any commercial fishermen who prior to the ban had harvested more than 400 kilos of eel a year. This seems to be a rather timid sort of a ban – but reading the ample collection of negative responses given to the Board of Fisheries from referees such as county administrations, fishermen, anglers' associations and private citizens, it becomes apparent how tricky it was for the Board to enact even this marginal restriction – applicable, strictly speaking, only to occasional eel fishermen, anglers and private citizens. Most respondents completely ignored the fact that eel reproduction had diminished by 99 per cent, and that the eel by all accounts was an acutely endangered species; instead they wrote page after page of the same old fisheries policies lingo, talking about minimum sizes, kilos, compensation, dispensations, dimensions of fishing equipment and the EU subsidiarity principle. A few were in favour though, but on the other hand did question why the ban only applied to recreational fishermen. One respondent concluded: "if the situation is judged to be so acute, a total ban ought to be the only solution."

Research on eels has barely begun. In 2006, a belated project was started to tag eels with data collection devices measuring depth and water temperature once a minute. Sixteen mature "silver eels" were tagged in the Kalmar Sound on Sweden's south-east coast, and eight of them were recaptured within twenty-four days. The results from this study have further confused and fascinated researchers into this mysterious fish. The data illuminated the eels' daily behaviour: it lies still on the bottom during the day, rising to the surface shortly after sundown and swimming steadily all night until just before sunrise, when it dives to find a spot on the bottom again to rest. Some eels had other habits though: one eel for instance swam a whole day, and at a greater depth than the others. Another stayed on the bottom one whole night. But the real news that researchers discovered was how eels during swimming regularly dive straight down into deep water. Sometimes to the bottom, sometimes just to a depth of 15–20 metres, to a level where there is a strong change of temperature (known as thermocline). Why these dives, once or twice an hour? One theory is that the eel is looking for deep water currents, helping it to orientate and find its way out of the Baltic Sea. Why it swims as close to the surface as it does is also not known, but one hypothesis is that, just like migratory birds, the eel has a compass instinct, perhaps even eyes sensitive to the earth's magnetic fields. It has been shown that birds "see" the direction of magnetic fields, though not in total darkness. All that is needed is a weak light source such as starlight for the eyes' magnetic sensor to function. The Board of Fisheries' eel expert Håkan Westerberg speculated in the Swedish fishing trade journal *Yrkesfiskaren* that eels possess the same magnetic compass as migrating birds.

But we just don't know. Nor do we know if restocked eels released in Swedish waters actually find their way out of the Baltic. All the eels in the experiment were recaptured along the south-east coast; none was found further south, or in the Öresund Strait, which is the only way out to the Atlantic – so questions about their ability to find their way to the Sargasso Sea are still unanswered (and are still in 2012!).

But why the dramatic decline in glass eels over the last decades?

Again, no one knows. One theory for their decline is that toxic polychlorinated biphenyls (PCBs) may have damaged the eels' reproductive capacity. Another theory is the Gulf Stream has become weaker and unable to transport as many eel larvae as before. The spread of eel herpes from eel farms in Denmark and Holland, now detected in wild eels, may not be a major cause of death, but has surely not helped the eels either, nor have hydropower turbines all across Europe that are estimated to account for around a sixth of the anthropogenic eel mortality.

The only unquestionable fact, established and recognised by more scientists, fishermen and bureaucrats than one would think would be needed – is that the eel is becoming extinct. And that the most obvious, imminent threat to the remaining eels is that they are still being fished, killed and eaten. Measures are of course also necessary to save eels from being massacred at hydropower stations in the rivers, but will not be of much help as long as unlimited harvesting of saved eels is permitted on the other side of the turbines.

Cooperation, dialogue and "balance" are the buzzwords in today's fisheries management. The dialogues takes place in conference rooms in Brussels, in Stockholm and all over Europe. It takes place between environmental organisations and fishermen, between scientists and fishermen, and is even formalised in the so-called Regional Advisory Councils (RACs), two-thirds composed of representatives of the fishing industry and one third NGOs such as consumer and environmental groups, jointly entitled to influence EU fishing policy – in an attempt to "balance" decision-making on fisheries.

A reasonable question is however, whether "balance" always is the best way forward, given the current acute state of world fish stocks? Is it really still in the common interest to try to reach consensus with an industry, when the activities of this industry are so obviously damaging the common good? As Henrik Svedäng, one of a number of frustrated scientists I talked to, put it: "The general view seems to be that we scientists are on one side and fishermen on the other and we are two stakeholders, in competition over the same interest, and that we need to compromise. But the truth is that it's the fishermen who have an economic interest to protect, not us. I've got nothing to gain by claiming that fish are being overfished; I get my salary anyway. There's a basic error in the way this is being perceived – as though you can equate fishermen and scientists. As scientists, we can't compromise with the truth! This is not a quarrel you can fix through mediation. We can't "balance" facts, it's not our job! The problem is not that scientists are making a fuss, rather that scientists haven't made enough fuss!"

Fisheries have long been very low on the news media agenda. This is perhaps the main reason that it has been so hard to interest politicians in taking strong action and so hopeless for all those with insight into the catastrophe to bring about change. Like a mantra, fisheries bureaucrats – themselves hostage to the industry because they are basically economically dependent on an industry to run – have clung to the notion of dialogue and consensus, and the hope of a change of attitude among commercial fishermen themselves.

But the question is: must a shift in attitude always be what precedes change? Isn't it true that it's often the other way around: that a change alters attitude?

There are many excellent examples of this: for instance the ban on smoking in restaurants or the introduction of traffic congestion charges. Both were presaged by loud protest and dire predictions of failure in Sweden. But only days into both the reforms, public opinion turned as if by magic. Even smokers found breathing easier in bars, and diehard motorists reluctantly recognised the advantages of less traffic in the city and the joy of hearing birdsong again. Everything improved and surprisingly few protested.

Should the world's politicians and fisheries administrators put their collective foot down and dare to push through comprehensive changes, the result would doubtless be similar. Ask fishermen in Newfoundland – they now all readily admit they were actually wrong in 1992! The fish wasn't an endless resource and they should have listened to the scientists!

Sometimes politicians have to take tough decisions, to get things right in a longer perspective than the next Election Day. When it comes to fisheries policies there are actually many good examples to follow. What is most needed is the courage to alter the perspective of fishery policy – from seeing fishing as an industry among others, to seeing it as a natural resources issue. And to realise that the world's fish resources are not the property of the fishing industry but of the world's citizens.

Let's begin here and see what can be done, and what, by extension, can bring not only healthier seas with more fish, but a more profitable fishing industry.

Marine biologists Daniel Pauly and Jay Maclean's book *In a Perfect Ocean: the State of Fisheries and Ecosystems in the North Atlantic Ocean* paints a picture of how the North Atlantic once looked, above and below the surface. The Pilgrim Fathers who arrived on the *Mayflower* at Cape Cod in 1620 were surrounded by whales as they approached land. The settlers on Chesapeake Bay reported not only whales but also large sea turtles and shoals of up to six-metre-long sturgeons that could capsize dinghies and canoes. The seabed teemed with lobsters, oysters and mussels; herring and sandeels roiled the waters. Salmon and sea bass were abundant and, further out, gigantic cod over two metres long patrolled the waters, quietly snacking on squid, sea cucumber and small fish. Here and there, shadows on the water surface betrayed giant bluefin tuna passing by at high speed, or a whale that filled the water with its clicking sounds and its song.

Daniel Pauly reminds us how important it is to recognise that a single human lifetime is too short for any of us to grasp the impact modern fishing has had on the oceans. Pauly calls this syndrome "shifting baselines"; reference points have moved, and fishermen today can at a stretch compare life at sea with how it was in the 1960s, the extent of living memory. But how would the settlers of Cape Cod take the news that the cod, the bountiful fish that gave the place its name, is now almost gone, as are the whales and turtles? How would the settlers of Chesapeake Bay feel to hear that the huge oyster colonies that covered the seabed and purified the bay waters, have decreased to a small fraction and that the bay has become an over-fertilised ditch?

We are all like the frog in Al Gore's film *An Inconvenient Truth*, the frog that never leaps from the cooking pot because it adapts to the slowly heating water. Swedish fishermen in their prime can still recall how much richer sea life once was. They remember porpoises, sharks and even catching loads of bluefin tuna in the Öresund Strait in the early 1960s, and every sport fisherman remembers how much bigger the catches once were – the length of an arm never suffices to show how big they were! But the reference points have been moved in such small steps over time that many have been able to convince themselves that what they are seeing are only natural variations in the catches; that some species come and go. And unfortunately, the layperson has not been able to follow the changes with his own eyes; no one has seen the transformation under the surface, from a sea simmering with life to one that is cleared and ploughed over several times a year by heavy beam trawls. No passers-by have seen species like cod, pollack, eel, ling, haddock, turbot, halibut, thornback ray, lamprey, basking shark, skate, tope, porbeagle, small-spotted catshark, Greenland shark, grenadier and spiny dogfish becoming increasingly rarer, until they finally wind up on the IUCN Red List of threatened species, and most of us have far too limited knowledge to even vaguely imagine what life would have been like under the surface of the North Sea or the Atlantic only a century ago.

But the underwater images I have seen of ploughed seabeds without any bottom animals or vegetation whatsoever seem to me to show the reference point of the end of time. Nothing can live here. No small fish can hide here. Here is the limit for how much we can harvest unpunished, what 19th century scientists considered an infinite resource.

Undeniably that limit has already been breached. We must start doing something. But what?

Establish no-take marine reserves!

There is actually something that can be done to restore some areas of the sea to a state resembling the pristine. Marine biologists talk about it, and many of them argue eloquently that at least thirty to forty per cent of the world's oceans should be decreed marine reserves with fishing banned.

A lot less than one per cent of the world's seas are today protected from fishing. The idea of terrestrial national parks and reserves with no exploitation of natural resources at all dates back more than a century – but has not won acceptance for the sea. The marine biologists who have pointed out the need to protect certain underwater environment from all exploitation, including fishing, have thus far been shouted down by the fishing industry. In Sweden, up until mid-2006 not a single no-fishing area was demarcated; not even in the first marine national park, Kosterhavet, had commercial fishing been banned.

Of course, fishing is not the only threat to fish stocks: pollution, eutrophication and climate change are also factors impossible to exclude.

Or are they? One way to ascertain whether it is fishing or other environmental factors that are threatening sea life is to establish control zones: marine reserves where fishing is banned. And indeed, the very few no-fishing zones around the world show astonishing differences to their surroundings. One of the better known is off Cape Canaveral in Florida, where space programme security means boats – including fishing boats – are not allowed. Nowhere else in the world can anglers catch fish as big as here – eighteen out of twenty world-record redfish have been caught right in the vicinity of the NASA's security zone.

After the explosion of the Challenger space shuttle in 1986 and the space programme had been shut down temporarily, the US marine biologists gained their first chance to examine the roughly thirty square kilometres off Kennedy Space Centre that had been off-limits to all fishermen since 1962. James Bohnsack, a marine biologist at Miami's National Fisheries Service told the *Los Angeles Times* (22-07-2002) that "the concentration of fish was unbelievable. I've never seen anything like it anywhere". The concentration of fish and marine life was up to twelve times higher than in the waters outside the security zone. Salmon bass, drumfish, robalo and brown trout were there in never-before-seen sizes. "Not so strange," Bohnsack commented. "These are fish that don't mature overnight. Salmon bass live until they're thirty-five, drumfish can reach seventy."

It is known that older and larger fish lay more and bigger eggs. Areas like this, as well as marine reserves established for other reasons, thus serve as "hatching areas" that produce a surfeit of fry, which also spread to adjacent waters.

US scientist Rod Fujita calls areas like this "insurances". If the fishery biologists who recommend fishing quotas by judging what are "maximum sustainable yields" of fish stock get it wrong, and the ecosystem flips, parts of the ecosystem are still untouched and have a chance to be saved for posterity. Fujita compares marine reserves to "money in the bank". No reasonable person bets all his capital on one card the way the world's fishing administrators and politicians are doing. Fujita observes that if the world's fisheries had been a multinational business corporation, then the CEO would have been fired long ago – for speculating away ninety per cent of corporate assets, and protecting less than one per cent in safe bonds.

Sweden is actually in the process of establishing six fishing-free zones: three in the Baltic and three in the North Sea. The first was set up in 2006: thirty-five square kilometres around the Baltic island of Gotska Sandön.

The decision was preceded by the Board of Fisheries' evaluation of 89 no-fishing zones across the world. The reports were overwhelmingly positive: after only two or three years miraculous changes were noted, with, on average, double the density of fish, three times greater biomass and much bigger individuals and greater variation of flora and fauna in the reserves compared with similar exterior areas. Some examples were even better, such as Goat Island Marine Reserve in New Zealand where the snapper are eight times larger than outside and the amount of fish is 14 times that in comparable areas. The Goat Island reserve, established in 1977 for research purposes, has had an unforeseen positive social spin-off – with school classes, families and outdoor lovers flocking there to dive and snorkel. The zone has become a "show room" for a living sea, a piece of nature that has created important new references for young people who have previously snorkelled in waters with very few fish, or who have seen fish only as rectangular packages in supermarket display counters.

There are also examples of how fishing restrictions, after a long period of overfishing, have not restored the sea. The most famous example is the Grand Banks in Canada where a cod fishing ban led to increases in shellfish, haddock and flatfish, but not in what was previously the dominant predatory fish: the cod. So there are no guarantees that ecosystems can be restored; nature might have found a new equilibrium where new species take over. This phenomenon is not necessarily unprofitable from the fishing industry perspective though. Some Canadian fishermen now make the same profits from harvesting shellfish as they did earlier from cod. But from an ecosystem perspective it is a catastrophe. Marine biologist Daniel Pauly calls it "fishing down the food web". After almost obliterating the largest marine mammals in the sea – the whales – mankind is hard at work overfishing the next largest creatures: the tuna, swordfish, shark, cod and salmon. Next in line are shellfish and pelagic fish, after that there will be nothing left to eat but "jellyfish sandwiches and plankton soup" in Pauly's words. Incidentally, commercial fishing for jellyfish already occurs in Australia and the US, which export the product to Japan. "Ten years ago, I said that as a joke. Now it's reality," observed Pauly in an interview in Swedish newspaper *Dagens Nyheter* (22-11-2006).

Would no-take zones in Sweden have an effect? It's hard to predict whether cod, for example, would recover under protection but some studies suggest it could. The usual argument against permanent no-take zones is that they would be ineffective if they are not extremely large, since fish roam over large areas. However, Swedish research has shown that there are several rather stationary cod stocks in the Kattegat, and the Swedish west-coast fjords Havstensfjorden and Gullmarsfjorden, that would benefit from a protected area. Tests in marking cod with tiny transmitters have shown that the cod moves territorially a lot less than previously thought, actually gathering in favourite locations that could be protected. Another study looked at two artificial reefs near Vinga Island in the Gothenburg archipelago. After fishing was banned there between 2003 and 2005, not only did lobsters increase 15-fold, but surprisingly, the cod abundance tripled. And individuals were far bigger than in other waters.

So there are reasons to hope that the six Swedish no-take zones will constitute "saving accounts" where fish will be secure for future generations, and that the "interest rate", the generous reproduction of fish fry, will spread its blessings outside of the zones and also benefit fishermen.

Ominous in the planning for the new marine reserves is the extremely defensive choice of the first Swedish no-take zone ever, the 35 square kilometres around Gotska Sandön Island. This area was chosen instead of 15 other alternative proposals of sea areas, with the weird motivation, to say the least, from the Board of Fisheries that fishing had already stopped there years ago anyway. "The principal reason for choosing Gotska Sandön is that it is the only area that has the potential to meaningfully demonstrate the management effects of a fishing ban, since the area has not been under fishing pressure at all for the last ten years."

Sound familiar? Ban eel fishing except by eel fishermen; impose fishing bans where fishing has already stopped? The Board's research had shown that the area had been rich cod fishing grounds until cod disappeared around 1986-1988. At that time, heavy turbot fishing took over, ending in its turn in 1995. Since then – no fish, and thus: no fishing. How the Board expected to see in a "meaningful" way the effects of a no-take zone after all the fish had already been gone for 15-20 years is difficult to understand.

Scientist Henrik Svedäng, involved in the decision himself, makes no bones that the choice of Gotska Sandön was not a choice made on biological considerations, but solely because the fishing industry no longer had any interests there. The choice, he says, was "a joke".

Discussions about marine reserves or marine protected areas often get confused, since definitions of what is a reserve or what is protection vary greatly around the world. In the EU, member states have committed to establishing marine "*Natura 2000*" areas, covering some percentage of Europe's seas. In the Nagoya biodiversity summit in Japan in 2010, the world committed to protecting 15 per cent of the world's oceans in marine reserves. This is all good – but the EU and the world have to be even bolder. Only no-take zones will give us the true "money in the bank" value that Fujita is talking about. Everything else is just going on with the same irresponsible speculation economy of the seas.

Ban bottom trawling!

In the 19th century, sailing ships and later paddle steamers began using the first kind of trawl net: a cone of netting dragged behind the boat. Older methods of fishing cod, such as longlines, trolling and bottom nets, in the new age of steam engines and industrialism, suddenly seemed hopelessly old-fashioned. The first so-called otter trawl was built in Scotland in 1892. Earlier, the trawl net was kept open with a beam but would work only if the seabed was smooth, whereas the new type had rollers or metal balls to help it across uneven seabed surfaces. It also had two heavy doors horizontally holding the net open on the bottom.

The otter trawl is still the model for modern bottom trawl nets though with better and stronger materials, and now dragged by increasingly powerful engines, at ever-increasing cost, with ever-larger dimensions, staying out for ever longer – to harvest fewer and fewer fish. Maybe it is time to re-think? In the long run, is this really the most efficient and economical way to fish?

A very interesting example worthy of the world's scrutiny is the Öresund Strait between Sweden and Denmark. For boat traffic safety reasons, trawling has been forbidden here since 1932. Both the Swedish and Danish sides of the strait are fished intensively though, but only with nets.

Oddly, in this small sea area affected by so much human activity potentially disturbing marine life – intensive agriculture, industries, the Barsebäck nuclear plant, heavy boat traffic and the newly built Öresund Bridge connecting the two countries – there are immensely more cod, haddock, whiting and lemon sole than in any other nearby waters. The size of individual fish is also much greater than in other Swedish or Danish waters. The figures are so obviously anomalous that no discussion ought to be necessary. Test trawling carried out throughout the 1990s in the strait brought up 500 individuals per hour sized over 50 centimetres; in the neighbouring Kattegat, the figure was less than 20 individuals. Catches of young cod were also strikingly much larger: 1,500 cod per hour, sized between 20 and 50 centimetres, compared with only 50 individuals in the Kattegat.

The explanation tendered by scientists for this wide discrepancy in catches should be familiar to us by now: this is what happens when the largest individuals – those best at reproducing – are spared.

Trawling is non-selective, scraping up everything from the seabed: wrong species, old car tyres, undersized fish and, if you're lucky, the odd whopper. Net fishing is much more selective; a specific mesh size catches a specific size of fish, the smaller ones swim through the mesh and the larger ones "bounce" off.

So why have these scientific results not attracted more attention? And why have commercial fishermen themselves not drawn any conclusions from them, and agreed on a voluntary ban on bottom trawling, or at least a trial?

Presumably because it's easier said than done. Force of habit, the tragedy of the commons and the risk that "if I don't do it, someone else will" are probably the best explanations. Besides, tax breaks on fuel make trawling for many hours for very few fish still profitable.

But a radical political decision banning bottom trawling would in all probability within a few years (after fish stocks had recovered) be greeted by the fishermen themselves the way the ban on smoking in restaurants was met by smokers: with joy and surprised gratitude that things had become so much better for all parties.

A significant step in the right direction was made in 2007 when over 20 Pacific Ocean countries agreed to ban all bottom trawling where sensitive areas were at risk. Thus, a quarter of the world's seas have been protected – from the Antarctic to the Equator and from Australia to South America.

Ban discards!

Trawling and unwanted by-catch go together hand in glove, which is easy to imagine. But most of us would have trouble guessing the volume of by-catch in an ordinary fishery. Normally anything between 20 and 80 per cent of what the trawl brings up is dumped back into the sea to become food for seagulls or just sink to the bottom – either because the fish are not commercially viable, or under the minimum landing size, or because the fishermen cannot land them because the quota is already filled. The ethics, economics and practicality of this practice has long been questioned, not least by fishermen themselves.

A fishing industry proposal for addressing the by-catch problem is for a "days-at-sea" system to replace the quota system. Fishing would then no longer be regulated by kilos and tonnes landed, but by time spent at sea. All the catch harvested on a sea day could be landed, regardless of quota, which would end a lot of discarding.

The system has been tested in different countries, with varying results. For instance in Alaska where they tried to limit the halibut fishery using the days at sea system; an experiment that ended in disarray. For years, the authorities chipped away at the number of sea days for halibut until they were down to just two days a year by 1994.

But despite the restrictions, no halibut fisherman had lost his dream of the giant catch. On the contrary, it virtually became a sport, with rows of boats equipped to the teeth having their engines running at midnight on the appointed day, to catch as much fish as was humanly possible in the allotted 48 hours. The "halibut race" resulted in a number of serious accidents among fishermen, lost equipment that continued "ghost fishing" for years, and a bizarre market situation where the year's entire halibut quota was snapped up by the processing industry and restaurants simultaneously. Market saturation meant the fishermen made less money, consumers had to make do with frozen halibut for the rest of the year, and halibut stocks did not recover. A race, in other words, with no winners.

A different system with individual quotas of halibut was introduced in 1995, ending the mad race. The season was extended to 245 days and 20 per cent of the boats were expropriated. Those that were left had a far easier time, since everybody had something to gain by protecting the halibut stock – one per cent of a large stock of fish is after all more than one per cent of a small stock.

The days-at-sea system was also trialled in the Kattegat in 2008. What was good about the system was the closer monitoring of how much fish was being killed, since fishermen were landing everything of value they harvested, not dumping it if wasn't part of the quota. Another advantage was that commercial fishermen no longer had the sense that they were always doing something wrong, a feeling that haunts them; if they are dumping quality cod because the quota has been surpassed they feel bad, but landing the cod is illegal!

Anyway, the results of the Swedish days-at-sea trial showed a clear tendency: fishermen were catching far more on sea days than on normal "quota days"– and especially cod. In fact, so much more that the experiment was had to be stopped very much in advance. The Board of Fisheries found that if all the boats in the Kattegat were to be allowed to fish freely on sea-days, the entire year's cod quota would be reached in only nine days.

In the Faroe Islands, where the days-at-sea system is used instead of quotas, the experience is the same – fishing is more intensive on those days since the fishermen have a greater profit motive. The Faroe authorities have thus combined the system with protected zones, where no fishing is allowed, and gear restrictions. But overall the experience of days-at-sea has been positive, especially since discards have dropped to a minimum, the entire catch is used, and all catch statistics tally – which is vital for marine biologists when calculating spawning biomass in the sea.

Days-at-sea may be a system worth trying, but trying it specifically in the Kattegat or other areas where the International Council for the Exploration of the Sea (ICES) has long recommended a total ban on demersal fishing – is far too risky. The stocks of cod and a number of other species of table fish are under so such pressure that scientists believe the only way to let them recover is total protection.

But the discussion about how to end discarding of fish must not be allowed to stand or fall with the days-at-sea system. Even with a quota system discards can be banned – fishermen can use more selective fishing methods to avoid unwanted catches in the first place, and secondly, they can actually avoid areas and times of the year when a lot of by-catches are expected. The phenomenon of "high grading" (that is, deliberately continuing to fish when your quota is already filled, and dumping smaller specimens to land only the selected bigger ones) must also be addressed. It has indeed been banned in EU, but is very difficult to control. That many fishermen are tempted is not strange when you look at fish auction prices: a "size one" cod (over seven kilos) fetches at least five times more per kilo than the smallest "size five" cod (under one kilo). When stocks are under pressure, and every skipper's quota is reduced to a point where he has difficulties making a living – high grading may be the only way for the skipper to make his fishing profitable – and this is undeniably the beginning of the end. In Canadian waters, the problem became acute in the years before the cod collapse off Newfoundland. In the book *In a Perfect Ocean: the State of Fisheries and Ecosystems in the North Atlantic Ocean*, a Canadian air force pilot describes what he has seen from the air:

> *One morning, we saw 40 or 50 Spanish pair trawlers working Green Bank ... Some seemed to have tails. When we came closer we saw it was dead fish. There must have been millions of fish in the wakes of boats that had just pulled in their trawls. They were sorting fish on deck and small haddock and other fish were being tossed over the side like confetti.*

Norway has a long-standing ban on dumping fish. Prohibition has helped bolster public trust in commercial fishermen as well as the fishermen's own pride and feeling that they are doing something right. And through the ban, the authorities can swiftly discover if catches contain too many young fish that should be left in the ocean. All this means that, if needed, temporary blanket fishing bans, known as real-time closures, can be called in areas where the concentration of young fish in catches is too high. In Norway, fish landed surplus to quota is sold, with the profits going to offset fishing inspection costs.

In 2011 the EU Commission actually proposed a discard ban, and it has been met with initial protest from commercial fishermen, but if it goes through it will ultimately bring them new pride and enjoyment when they see the advantages in showing consumers fair practice. A pinch of political courage – nothing more – is needed if the daily, meaningless fish massacre in our seas is to be halted!

Change the perspective

Another alternative to the current quota system based on free access – is what are usually called individual transferable quotas, ITQs (but they can also be referred to as for example ITRs; Individual Transferable Rights, or TFCs; Transferable Fishing Concessions). For instance in Australia, New Zealand and Iceland, fish quotas are commodities that can be owned, and can therefore also be traded. The advantages of the ITQ-system are that quota owners can plan ahead and do not benefit from short-term overfishing. They can regard fish stocks much like farmers see their fields – as something that has to be cared for, diligently managed, and harvested only when the crops are ripe, in order to make it as valuable as possible the day the harvest is sold. Australia and New Zealand apply the system so that fishing industry fees pay for fishery administration and fishery research. Catch limits recommended by scientists are thus not seen as impositions but as reasonable advice to be followed if the fish population is to be kept at a good, high and ecologically sustainable level. Not infrequently does the industry actually turn to the management authorities to ask them to *lower* the annual total allowable catches – an unthinkable scenario in a pure quota management system.

The system of ITQs has in some cases demonstrated its suitability with regard to the protection of fish, although less successfully if regional policies are taken into account. In Iceland especially, people are upset that ITQs have been bought up by ever-larger companies and ultimately concentrated in a few hands.

ITQs were introduced in Sweden in 2009 for the so-called pelagic ships that trawl for herring and sprat. More than eighty boats have been attributed quotas, which has quickly led to many small herring trawlers quitting the business, though not completely (many have switched to fishing nephrops, which has actually increased pressure on bottom-living species).

Potentially, the ITQ system has many benefits. Perhaps the main one is that it provides a new perspective on fish as a valuable resource that is not "for free", laying the foundation for the paradigm shift needed in fishing. We cannot any longer hold on the old dream of free fishermen out on the free seas where everything is free and inexhaustible. Technology has taken us far beyond the time when that was possible. Today even the small-scale fishermen are far too effective to be allowed to operate without restrictions and the seas can no longer be "free for all". And even if ITQs are not a perfect answer, one thing is certain: the system we have today which has precipitated a catastrophic decline in fish populations and the fishing industry's economy, besides costing taxpayers billions and billions of euros and kronor – is not the answer either.

Still – no system must ignore that, fundamentally, the fish belong to us all. Those who exploit fish must pay for them and take responsibility for them. The obvious disadvantage with the ITQ system is that the state often is too casual in giving away fish to private interests, only based on one criterion: historical catches. In other words, the biggest quotas are allocated to those that have overfished most in the past, which is a rather perverse incentive. As far as Sweden is concerned, gigantic loss-making boats like *Tor-ön* and *Torland* that have never contributed a single tax krona to the state, have received generous quotas from the same state in the ITQ allocation. Together, these two vessels received quotas of 10,000 tonnes of herring and sprat, worth approximately 1.5 million euros a year. Will this make the boats more profitable? Does it benefit society and the seas that these particular boats get such a large allocation?

In Denmark, which introduced ITQs in the pelagic segment in 2006, the eight biggest "quota barons" own quotas estimated at 4.8 billion DKR (644 million euros). The biggest boat, 67-metre-long trawler *Isafold* from Hirtshals, once bought for 17.5 million euros, is now worth an astonishing 134 million euros, according to estimates done in 2011.

Another way to design the so-called rights-based management is to link the ITQs to certain sustainability criteria. Then those who cannot meet minimum environmental and/or social criteria will not be eligible to receive quotas, while those who score well (who fish responsibly, legally, using little fuel, harvesting selectively and for human consumption, and who also contribute to local and regional employment – and pay taxes!) – gain priority access to fish.

What is crucial is that the ITQs are not casually distributed and that they can be regularly reviewed, and withdrawn if the rights' holder has not performed well. If the privilege to fish is abused, then society has to regain control. Because the sea is still ours – yours and mine.

Stricter control, stiffer penalties

In 2011, almost 20 per cent of the world's catches were estimated to be illegal, making IUU (Illegal, Unregulated and Unreported) fisheries the number two fish producer in the world (after China), with an estimated annual value of 23.5 billion USD.

Poaching is a global problem, and the difficulties in dealing with it appear with tragic clarity in Bruce Knecht's exciting book *Hooked: Pirates, Poaching and the Perfect Fish* (2006), an account of how the Australian and South African coastguards, after a hunt lasting more than a month, together chased down a fishing boat with a haul of the threatened deep-water Patagonian toothfish valued at some million euros. After gathering evidence that the fish had been poached in Australian waters, the Uruguayan vessel *Viarsa* was escorted to Australia and a lengthy court case ensued.

The *Viarsa*'s owner, Spanish multi-millionaire Antonio Vidal Pego, hired Australia's best legal team to do its best to sow doubt among the jury. The official Uruguayan government fishery observer who was on board the *Viarsa* when the Australians boarded the ship refused to attend the trial, all of a sudden claiming he had been seasick for the entire voyage and was unsure whether he had been in the Atlantic, the Indian Ocean, or the Antarctic Ocean for that matter. This is how one of the world's most blatant incidents of poaching and one of the extremely few that had resulted in discovery, boarding, arrest and in fact trial, still ended in acquittal in September 2005.

Globally the problem of combating illegal fisheries seems hugely challenging, considering the fact that pirates with very lucrative incomes are to be controlled by coastguards in developing countries or fishery observers with very modest incomes.

Cross-border cooperation and serious international pressure are needed if the poachers are to be stopped. There are already blacklists of suspected poacher boats that should prevent them from entering any harbours, and an IUU-regulation recently came into force in the EU, that requires every fishing vessel that enters into the EU to have sufficient catch documentation – but offenders can too easily sidestep this by trans-shipping, or landing their catch in countries with laxer controls and high degrees of corruption. On an international level there is a lot left to be done to come anywhere near addressing this problem that is contributing to emptying the waters of developing nations, as well as the high seas (international waters) – and also – EU waters.

Of course not all EU fishermen are innocent when it comes to IUU. In the Mediterranean the extensive illegal bluefin tuna fishery has been debated in the media at length. In Sweden illegal fishing amounted to around 20 per cent of the legal catch, according to a study by the European Commission in 2007. This estimate was arrived at by checking the catch differences between inspected and uninspected boats. On average, Swedish ships that underwent inspection reported 20 per cent higher catches than those that were not inspected.

Illegal fisheries is a high profit, low risk activity in all seas, Sweden not being an exception. Here very few infringements are discovered and proved, and even fewer punished. The EU has even criticised Sweden for its low ratio of fishing transgressions that lead to sanctions: only 17 per cent in Sweden compared to 82 per cent in the rest of the EU. In addition, Sweden has very light penalties; the average fine was only around 1.350 euros (1995 – 2001), compared to the UK where it was 10.000 euros, and in Ireland almost 15.000. Many Swedish fishing crimes – and not only minor ones – also outlive the period of limitation and never come to trial. For example one case that concerned two boats that landed mackerel worth 30.000 euros in a banned time frame and did not record it in their logs was never brought to trial. All in all, the risk of discovery and/or punishment is so small that certain fishermen coolly risk breaking the rules; the tiny chance of being fined is only added to the regular operating costs.

To deal with the cheating, first of all, better monitoring of fishing boats is needed. Vessel monitoring system (VMS) transmitters that show a boat's exact position should be standard on practically every fishing boat in the world, as too should electronic log books trans-mitting in real time the catches and activities of boats. A global register of fishing vessels should be established as soon as possible, and transnational IUU crimes should be dealt with on an international level, also considering the vast amounts of crimes connected with IUU fisheries. It is not only a conservation crime, or a breach of fisheries management rules – IUU can also be tax evasion, money laundering, bribery, corruption, false declaration, customs offences, even human trafficking and organised crime – very many severe crimes that should be treated accordingly.

Penalties should thus be raised and harmonised in the entire EU for a start. Prison terms, revoked fishing and boat-operating licences are painful penalties that would doubtless deter those commercial fishermen who break the rules because "everybody else does it" and because the law does not look on it severely enough.

Poaching is an international crime with annual takings worth billions of dollars or euros. There is no figure for the world total of fishing fines, but if they correspond to the EU scale they are a couple of thousandths of the value of landed illegal catches.

The pirates have much to gain and little to lose.

Improved traceability and consumer information!

Dumping the responsibility for systematic legal overfishing on the consumer is in my view nothing but political cowardice and haplessness! Legally caught fish (eel, bluefin tuna or cod) that never should be harvested in the first place are now sold in every shop and consumers are encouraged by politicians to "exert influence" by choosing to buy or not. But based on what information? How can even the most environmentally conscious shopper know which fish to choose? Labelling of origin is often so poor that it scarcely deserves the name. According to EU rules, fish does not have to be labelled more precisely than with the species scientific (or 'Latin') name and the FAO's division of the world seas. Fish caught in the Barents Sea, or the Öresund Strait, or the Kattegat, or the Bay of Biscay can all be marked "Fishing zone 27, North-east Atlantic" – an area of 13 million square kilometres! The informed consumer who knows that cod from the Öresund population is in better condition than the Kattegat cod can deduce nothing from such labelling.

Then of course, rules of origin do not apply to the large market for processed fish. On packets of fish fingers, fish au gratin or battered fish portions you will find completely meaningless information such as that the fish comes from "the cold, clean waters of the sea" (Findus), or that the fish fingers contain "60 per cent fish" (Eldorado).

A responsible fishery needs complete transparency towards consumers; demands on the industry must be tightened. Correct information on species and harvest area, according to the far more detailed sea demarcation proposed by the International Council for the Exploration of the Seas (ICES), are obvious demands. So too is information on harvesting methods and the names of the vessels involved. Responsible and honest fishermen would surely welcome the opportunity to deliver their "own" fish all the way from the net to the consumer. Eco-labelling of some Swedish herring and shrimp boats is an excellent example of this.

And more detailed labelling will also help fish merchants to be able to inform consumers on where the fish comes from, and consumers who in turn will be able to significantly influence the state of the seas by refusing to buy what we know are threatened stocks. Unfortunately, this consumer weapon is currently stymied because the level of consciousness among consumers in the EU member states varies considerably, so that fish is harvested regardless of whether local consumers boycott it or not. For instance, when Swedes and Swedish retailers decline North Atlantic cod or cod from the Kattegat, the fish is not saved but exported to countries in southern Europe instead.

Guarantee independence of research!

Research into sea life must develop. Sometimes it must collaborate with the fishing industry, and sometimes it must be policy-driven, but overall it should also be largely independent from fisheries management. Much of current marine research is in a rut, serving the fishing industry with information and expending more energy and research money on catch charts and selectivity in fishing equipment than on fundamental knowledge such as researching animal plankton, eel reproduction, the lifecycle of the cod, or diseases in wild fish. This limits radical new approaches. Research on fish stocks should be integrated into a wider ecosystem approach; fish should be dealt with not only as "food stuff", but as fundamental factors in the world's marine ecosystems.

A new way of looking at fish in the sea also demands a new organisation on land. Renaming the General-Directorate that deals with fisheries issues in the EU from DG Fish to DG Mare is a step in the right direction. But in most EU member states fisheries, and the research units tied to it, are under the agricultural minister, and not the environmental minister. Risks are that state employed fisheries scientists feel the political pressure to produce results that do not reduce fishing possibilities. A large number of people in research and public administration are still stuck in old structures and mindsets and dependent on income from their jobs. They are thus inclined to favour the status quo in fisheries policy.

But it is time to recognise that there are a lot of stakeholders in the sea and its resources, and with the so-called marine spatial planning (a method of mapping all the combined interests and impacts humans have on the sea) it will be obvious that we need objective, independent science to evaluate how we can best use the sea, and at what risks.

These issues must be considered at the highest political level. What do we need research for? What do we need the seas for?

Dump the annual quota system!

The annual quota horse-trading in the Council of Ministers thus far has had disastrous, sometimes absurd consequences. For Sweden it has meant in practice that for many years there was no limits on cod fishing at all, since fishermen were not even able to catch enough to meet the agreed quota. In 2003 and 2004, Sweden's quota was pegged at 17,000 tonnes, but total harvest was just over 14,000 tonnes. In 2005, the quota was reduced to 14,000 tonnes but only 10,000 were caught. The 2006 quota was kept at the same level but catches were again considerably lower: around 11,000 tonnes.

Now the Baltic cod is managed through a long-term management plan which does not allow for changes of yearly quotas by more or less than +/- 15 per cent compared to the previous year's quota, and this seems to be a good model – at least in a scenario where the cod has started to recover. Without the plan, quotas would doubtlessly have been set much higher in the last two years, since there is, in a one-year perspective – room for further increase in catches, without risk of stock collapse.

Long-term management plans, together with the goal of achieving MSY (Maximum Sustainable Yield) by 2015, is thus a strategy already being introduced by the EU, intended for all fish stocks and sea regions.

This is indeed a welcome approach. Politicians will now have to ponder the state of the sea and its fish in 10 or 20 years, and decisions will have to be made that go beyond annual short-termism (and governments' attendant wheeling and dealing to get the highest possible annual quotas for their own fishermen).

If the EU in its imminent reform of the Common Fisheries Policy does not continue with the implementation of long-term goals, the sea will probably never get a chance to recover. Short-term economic gain will always trump vaguer environmental losses and the downward trend that has so far emptied the sea of between 70 and 90 per cent of large predator fish will continue.

The surest way to achieve the long-term goals of sustainability would be for the EU to look closely at the US legislation, where the Magnuson Stevenson Act has actually made overfishing illegal. According to the 2007 act, managers are obliged to achieve good and sustainable status for all commercially fished stocks by 2011. This obligation relieves legislators from the burden and temptation of compromising scientific advice every year when quotas are negotiated – overfishing is actually against the law!

So, finally, one crucial question remains: if European politicians do not manage to take the responsibility, how can we as citizens put an end to this; the most meticulously managed, subsidised, long-lived, well-documented and closely monitored environmental disaster of all times? It seems meaningless to demand more research: the data is already there. It seems meaningless to inform those in power: they already know. Meaningless even to boycott: market laws have been rendered impotent by forces beyond our control.

All we can do now is protest, in the ways we can, in whatever forums we can, tirelessly and loudly, for our own sake and that of coming genera-tions, for the sake of democracy and common sense, for the sake of the planet and all its marine species.

Because the fish does not have a voice.

And the sea – is silent.

MEASURES TO BRING BACK LIVING SEAS

● **Establish marine reserves with fishing bans!** No-fishing zones let the fish grow back and have been shown to be beneficial also for adjacent waters. No-take marine reserves should be as accepted as no-hunting natural reserves on land.

● **Ban discarding of fish!** And temporarily prohibit fishing in places where there are large concentrations of young fish! The obligation to land all fish and count it off the quota will end the immoral practice of throwing away edible fish; it will help fish stocks to recover and make catch statistics and stock assessments more reliable.

● **Introduce mandatory environmental impact assessments on fishing operations!** On land, companies planning operations that can impact wildlife and the environment have to seek permission and submit environmental impact assessments. Fishing is the main cause of death for fish, and has a huge impact on marine ecosystems, so fishing methods should be evaluated before given a go-ahead, and best practice principles applied.

● **Limit bottom trawling!** Experience from the Öresund Strait where only net fishing is permitted shows that demersal stocks are in far better shape there than elsewhere. Seen over time, the economic efficiency of bottom trawling must be questioned. For every fishery the most sustainable fishing method must be applied.

● **Combat illegal fisheries.** Electronic logbooks and satellite monitoring should be mandatory, even for small boats. Raise penalties for infringements; today, it often pays to budget for these. Publish offenders' names on an international blacklist accessible via the web. Improve working conditions and salaries for coastguards in all member states. Prioritise international cooperation to prevent illegal fishing worldwide, and use available techniques to make all fish traceable from the plate back to the net.

● **Improve origin labelling.** Consumers must be given an honest chance to see what fish products contain, who has harvested the fish, with which methods, and where the fish is from. The industry must have the opportunity to show that their catch comes from environmentally sustainable populations. Voluntary initiatives from the industry are of course welcome; compulsory regulations are even better.

● **Implement long-term management plans.** End the yearly quota quarrel, set up long-term objectives, and follow the scientific advice. Follow the US example and make overfishing illegal! This would relieve the politicians from the burden of making quota decisions that are unpopular among fishermen in the short term. Stocks should always be kept at levels above MSY (Maximum Sustainable Yield), with considerations taken of their ecosystem role.

● **Integrate fisheries in a wider marine policy.** All activities that affect the marine environment should be considered in an integrated manner. By planning all marine activities together wise priorities are easier to make.

● **Rethink who are stakeholders!** The main stakeholders are not the few thousand people who earn money from fishing but the general public and future generations: they have a right to sustainable fish stocks and living seas. There must be an end to the subsidies that allow more people to stay in the trade than there are fish to sustain them. The necessary reductions to the fishing fleets need to done, starting by eliminating those who fish in the most environmentally damaging way. Limited, individual or community-based, fishing rights could be a way forward.

FISHING GLOSSARY

Fishing terminology is notoriously arcane and full of acronyms. Here is a brief aid to those who want to do further research.

Adipose fin clipping	marking of released farmed salmon by removal of the little fatty (adipose) fins on their back
Aquaculture	fish farming
By-catch	catches that were not the primary target
BACOMA trawl	selective trawl, named after the Baltic Cod Management
BFT	bluefin tuna
B_{pa}	biomass, precautionary approach. Lower limit for population size according to the precautionary approach
B_{lim}	biomass limit for population; size under which the population is at acute risk of collapse
BSAP	Baltic Sea Action Plan; plan to restore the good ecological status in the Baltic by 2021

CITES	the Convention on International Trade in Endangered Species of Wild Fauna and Flora
CCAMLR	Commission for the Conservation of Antarctic Marine Living Resources, see RFMO
CFP	EU Common Fisheries Policy
Demersal fishing	fishing of species who live near the sea floor
COFI	Committee on Fisheries of the FAO, see FAO
DG Mare	Directorate-General for Maritime Affairs and Fisheries; the EU Commissions agency, previously DG Fish
Discard	unwanted by-catch thrown back into the sea
EEZ	Exclusive Economic Zone; 200 nautical miles zone where the coastal states have the right to marine resources
Ecosystem approach	the view that all marine life is interconnected
EFF	European Fisheries Fund
EIA	Environmental Impact Assessment
Europêche	association of national organisations of fishery enterprises in the EU

FAO	UN Food and Agriculture Organisation
FAO CCRF	FAO Code of Conduct of Responsible fisheries, adopted in 1995
F	Fishing mortality rate, meaning the share of a fish stock that dies through fishing each year
F_{msy}	maximum yearly share of a fish stock that can be killed through fishing if stock is to be kept at MSY level (see MSY)
F_{pa}	maximum share of fish stock that can be killed through fishing according to the precautionary approach
F_{lim}	maximum share of fish that can be killed through fishing, which if surpassed leads to risk of stock collapse
Fishing effort	fleet or boat capacity multiplied by days at sea; in general, any quantified measure of the gear and/or activity deployed to catch fish
FPA	Fisheries Partnership Agreement; EU fishery treaty with a developing country

GES	Good Environmental Status. The EU seas should have GES according to certain criteria established in the MSFD, by 2020.
GFCM	General Fisheries Commission of the Mediterranean
GT	gross tonnage, gross capacity; measure of fishing fleet capacity
High grading	dumping of smaller fish so the quota can be filled with more profitable larger fish (not allowed in the EU)
ICCAT	International Commission of the Conservation of Atlantic Tuna
ICES	International Council for the Exploration of the Sea
IMP	Integrated Maritime Policy, the EU policy aiming at coordinating decision-making on sea issues
Industrial fishing	pelagic trawling, for instance for sprat or blue whiting, to produce animal feed
IOTC	Indian Ocean Tuna Commission, see RFMO
ITQ	Individual Transferable Quota
IUU	Illegal Unreported Unregulated fishing; poaching
Landing	the fish that is landed
M74	salmon disease caused by thiamine deficiency

MSFD	The Marine Strategy Framework Directive; the environmental pillar of the IMP. See also GES.
MPA	Marine Protected Area
MSY	Maximum Sustainable Yield, the share of a fish stock that could be harvested each year without stock depletion. All countries have committed to apply MSY by 2015
NAFO	North-west Atlantic Fisheries Organisation, see RFMO
NEAFC	North-east Atlantic Fisheries Commission, see RFMO
NGO	non-governmental organisations, such as WWF, Greenpeace, Oceana, etc
NOAA	National Oceanic and Atmospheric Administration, the US federal agency dealing with the oceans
Pelagic fishing	trawling in the water column between the seabed and the surface
RAC	Regional Advisory Council
Recruitment	the annual increment of young fish to the stock
Resource	fish

RFMO	Regional Fisheries Management Organisation; intergovernmental organisation tasked with regulating fisheries of highly migratory species and fisheries outside EEZs, see for example ICCAT
SAP	Salmon Action Plan
Sea days	or "days-at-sea"; a method of limiting fishing effort by only allowing fishing on a certain number of days
Selectivity grid	grid to filter out unwanted fish
SFR	Swedish Fishermen's Association
Smolt	young salmon migrating from fresh to salt water
SSB	Spawning Stock Biomass; the part of the population at sexual maturity
Stock	part of a species, usually living in a distinct place; species can have many stocks, or populations
TAC	Total Allowable Catch (per year)
Target species	the type of fish aimed for
Undersized fish	fish under legally permitted minimum dimensions
UNCLOS	United Nations Convention on the Law of the Sea, adopted in 1982
UNFSA	United Nations Fish Stock Agreement, adopted in 1995

UNGA	United Nations General Assembly
WCPFC	Western and Central Pacific Fisheries Commission, see RFMO

MEDIA Q AND AS

Using effective lobbying, the fishing industry has been able to feed the media with a few seemingly feasible arguments. The four commonest are:

"Cod is not at all a threatened species."

Response: True, as a species the cod (*Gadus morhua*) is not threatened. Even if some populations, or stocks, are completely or partly wiped out, the cod species will surely survive somewhere. But locally, the cod is definitely threatened. The world's largest cod stock, off Newfoundland and Labrador, in eastern Canada, collapsed in 1992 and has not recovered. In the North Sea some stocks are gone, and others are on the verge of collapse. For each stock that disappears humanity loses another important source of protein, and the sea's ecosystems lose their resilience towards climate change and other environmental impacts.

"The scientists are wrong. The fishermen see that there's lots of fish, and they know best."

Response: Fishermen can find large aggregations of fish, maybe even *the* last aggregation of a population, even though a stock as a whole is on the verge of collapse. Modern fishermen have access to sophisticated sonar and other equipment that effectively leads them to wherever the fish are. Scientists, on the other hand, do survey trawling based on grids in the sea that are checked year after year, and they can compare to more than a hundred years of records.

Fishermen sometimes accuse scientists of fishing with outdated equipment saying they are incompetent. But the whole idea of scientific stock assessments is to use the same gear year after year so that you can compare the amount of fish caught under the same conditions. They also use data from fishermen's logs from which they can see that catch volume is taking more effort (more hours at sea) than in the past.

"It is perfectly legal to buy legally caught fish, and the stores are full of them."

Response: Yes, a lot of fish is caught legally – but this does not imply it has been caught in a sustainable or ethical way. Every eel, Kattegat cod or bluefin tuna legally caught still represents a gamble on the part of politicians who, with scientists' warnings ringing in their ears, gamble with the survival of species for the sake of a few jobs.

"Why don't they stop fishing if the situation is that bad? Commercial fishermen are those who stand to lose the most if fish were to disappear."

Response: Since fish are a free and motile resource in the wild, it is hard even for single national governments to protect stocks from being caught – and a single fisherman who voluntarily fishes less will only see others gain from his sacrifice. The "if-I-don't-others-will" philosophy, in combination with the boom in EU funding in the 1990s that encouraged fishermen to increase the size of their boats and hence their bank loans, means many see no alternative other than to keep on fishing and hope for the best. Many fishermen also believe that if some species are depleted, then other species will come instead that will compensate for the lost fishery – as when sprat and herring populations exploded in the Baltic due to the decline of the cod – or when shellfish came instead of the cod on the Grand Banks in Canada.

TWO SECONDS

One of the most vertiginous analogies I have ever come across, which I've read a number of times, (most recently in Sylvia A. Earle's book *Sea Change – a Message of the Oceans)*, goes like this: imagine that we could condense the Earth's 4.6 billion year history into just one year and then watch it unfold on a movie screen, starting the 1st of January. For eight long months we would sit in the dark witnessing asteroid crashes, erupting volcanoes and all those strange chemical processes and coincidences that result in the planet *Tellus* evolving into a blue, water-covered exception in a dry and cold universe where both life and water are exceedingly rare phenomena. Not until August would we see unicellular organisms – life! – appear for the first time. The marine invertebrates would not make their debut until November, and dinosaurs and mammals would only materialise on the screen one week before Christmas – but the giant lizards would of course disappear again some days before New Year's Eve.

Our own species, *Homo sapiens,* appears around 23.45 the 31st of December. And modern society flashes by during the film's very last two seconds.

When I began writing this book in 2003, I never could have imagined the media headlines that suddenly would become so frequent during the winter 2006/2007. After the Stern report was published, heads of government and journalists suddenly for the first time seemed to take the threat of climate change really seriously. The "on-the-one-hand" and "then-on-the-other-hand" type of reporting that had haunted climate change media coverage for such a long time all of a sudden evaporated into thin air, with newspapers of all colours producing almost endless in-depth coverage on the why and the how rising temperatures were changing the face of our planet. All of a sudden the future seemed surrealistically uncertain. There was serious debate about whether Nordic skiing events would have to move farther north! And about how northern Sweden and Siberia would become agricultural bread baskets when the European mainland would turn into desert. In one of its climate reports, the United Nations said limiting the temperature rise to 2°C would be a triumph for global society – and if nothing was done the increase could be as high as 6°C or even 8°C, radically changing life on Earth in completely unpredictable ways. The news about a global seed collection project for safe storage on the cold island of Svalbard reinforced my surrealistic feeling – imagine that I would experience all this, just during the brief blink in our planet's history that is my lifetime!

But I am not really surprised. None of us can be. We knew about the madness while it was going on, but basically we just ignored it. None of us have been blind to it; we have actually all commented on it numerous times: look at all the people just sitting there in traffic jams; look at all the unnecessary Christmas gifts everyone buys; look at the waste of energy, the irresponsible release of toxic pollutants, the brainwashing effect of advertising, the fashion industry that deceives us into buying new clothes every season; look at finance speculations, those strange stock market reports in every news bulletin; look at the way we mistreat animals and how we abuse nature. We all know about the destructive madness; we are not blind to it. We understand what we are doing while we're doing it, and our only sorry excuse for doing it is that everyone else is doing it too.

This behaviour doubtless has evolutionary advantages in terms of survival of the human species: our instinct to follow the herd, not to challenge authority and to accept things as they are. But now, after a comparatively very brief moment in the history of our planet, it is becoming increasingly clear that this instinct of our species can also lead to catastrophe. And that the only way to avoid this is to start asking really fundamental questions about the way we live our lives: not only for the planet's sake, but also for our own sakes.

American professor of history Theodore Roszak gives an interesting perspective on *Homo sapiens'* fundamental sense of dissatisfaction in modern society in his book *The Voice of the Earth*.

Roszak is the founder of 'ecopsychology', a psychological theory based on the notion that growing mental ill-health is due to us having become detached from our natural roots; our oneness with nature – the nature to which we ourselves belong.

Roszak tracks the mental shift that made it psychologically possible for us to abuse and plunder the planet to the time of the Enlightenment and the development of the natural sciences. Only when we no longer saw nature as God's creation or a place where every tree, flower and animal held magical powers the conditions for our industrial and consumer society were created. Around the time of the Enlightenment, Roszak writes, we started seeing ourselves as more and more independent from God and also as separate from nature, a view that was confirmed by the birth of modern psychology. The sexually repressed bourgeoisie of the turn of the century Vienna were the subjects of Sigmund Freud studies and their thoughts and dreams formed the basis of his theories – theories that still influence the way we see ourselves today.

In Freud's writings there is absolutely no mention of man's need for contact with nature; on the contrary, animals, "savages", sexuality and even women (!) were all symbols of dark, uncontrollable forces threatening civilised man. Psychology set the modern man on an introspective path that further heightened his feelings of loneliness and futility in a universe to which he no longer felt connected. We forgot, Roszak says, that our lives – our very existence – are irrevocably interconnected with the planets.

This "forgetfulness" (although most of us haven't forgotten at all, only followed the herd!) has been conditioning our behaviour for more than a century now: a blink of an eye in the context of the history of the universe, but long enough for us to disrupt the entire balance of the planet. We have overfished the oceans, cut down the forests and poisoned the air and the lakes – and we have not enjoyed it. None of us has really enjoyed it.

Roszak is convinced that we, the modern people, would feel so much better if we re-embraced nature and drew solace from it; if we devoted our lives to repairing it, caring for it and enjoying it.

Ecopsychology has many followers. Various forms of nature therapy have proven more effective than so-called happiness pills, SSRIs (selective serotonin re-uptake inhibitor antidepressants) – or years on the psychoanalyst's couch. Cardiac patients with a picture of a landscape on the wall of their hospital room recover faster than those with an abstract motif to look at. Therapeutic gardens lower people's blood pressure and are unrivalled at healing stress-related symptoms. Elderly people live longer if they have a household pet. We have all experienced how problems diminish when we see a beautiful landscape, or we pat an affectionate dog, study the miraculous beauty of a flower, see an unusual bird, or just watch the waves glittering on the sea. Nature is us; we are nature. And we need it.

But does nature really need us? It may be time to start asking ourselves also this odd question. Will our contribution to the complicated web of life that is Earth, be that we exerted our power only to destroy it? Roszak, like James Lovelock, founder of the Gaia theory – is unafraid to ask the hard questions. What is man's evolutionary significance, in particular the role of human consciousness?

If we acknowledge ourselves as part of nature, parts of the life-filled globe that astronauts intuitively sense is a single life form, if we acknowledge our dependence on water, oxygen, vitamins, proteins, love, social interaction, sunshine, beauty – the entire unfathomably complex system that is life on Earth and that Lovelock calls Gaia (meaning that the entire planet can be seen as a single biologically connected system in which every part is equally important to the whole) – then what is our role?

This is the question we should perhaps be asking today. As the former US Vice President Al Gore wrote in his book *An Inconvenient Truth*: this is a unique moment in history. We have finally reached the end of the road, and old values must be cast aside. Gore states of course the point that some of us have heard many times already, but which is still worth contemplating: that the word "crisis" in Chinese consists of two characters, one meaning danger, the other – opportunity.

I find another beautiful word in another book: "physis". In his book *Heal the Ocean – Solutions for Saving our Seas*, marine biologist Rod Fujita uses the word to describe the extraordinary self-healing powers that he has recorded in areas of ocean that humanity has left in peace for a while. Marine reserves that are protected from fishing are few and far between round the world, but where they do exist they have become miniature underwater paradises sooner than anyone dared hope, attracting and dazzling divers and snorkellers with their natural beauty and biodiversity.

Physis is a Greek philosophical and scientific term to describe the miraculous intrinsic power that allows children or plants to grow, or wounds to heal. But how does this fit in with other universal laws of physics, such as the entropy law: the law that says that all energy is constant but disperses? In Jeremy Rifkin's book *Entropy: A New World View* he explains the principles of thermodynamics whereby all energy in the universe eventually converts from orderly to disorderly forms, and ever since I read this in the beginning of the 1980s I have been unable to see how we humans do anything other than work our socks off to hasten the law of entropy and allow disorder to succeed. We are extracting all the earth's natural resources, burning its oil, allowing soil to erode and ice to melt. The law of entropy, also known as the second principle of thermodynamics, chimes with our view of the world as a place to conquer, exploit and consume.

But what of physis, the force that allows a tiny seed to germinate and a leaf to decompose into earth and give new life to a flower? The force that sweeps away pollutants, covers an abandoned car with vegetation until it is out of sight, the force that inside a stem cell decides whether it will develop into a foot or a brain, that allows a microscopic sperm from a man and an egg from a woman to develop into a child – a sentient human being that can rejoice in the sight of a dolphin, feel empathy for the smallest life forms on the planet, and can read these words and pause in self-reflection?

Physis is the miracle of life. I mean this without any religious overtones other than that failing to see it would be tantamount to losing our belief in life. The absence of this belief makes it instead too easy for us to join the service of death: vacuuming the seas of all life, ravaging forests and spewing out toxins, contributing to the destruction of the planet because we have become blind to its beauty. Because we, as a consequence of the shortcomings of our species, have become obsessed by a nothing but frustrating culture of mass consumption in which none of us will see any point when we lie on our deathbeds ... because we have kidded ourselves that happiness lies in the things we can buy.

When in reality the greatest happiness is for free.

Physis is the magical phenomenon that is not possible to isolate, because it vanishes as soon as you start reducing nature to its smallest elements. It is what binds cell nuclei, quarks and atoms into things that work. It is what makes the sum of the parts larger than the parts individually, it is the power of ecosystems, of which we ourselves are part. This is the teaching of Lovelock and Roszak and the view of every marine biologist, artist or ordinary person who has ever felt the wonder of life on Earth.

I was cleaning out my basement not long ago. As someone who finds it difficult to throw things away, I was delighted to find an old metal container with bits of rubbish still inside – a piece of paper, a ticket to a football match and a bottle top. My reaction was to throw it all away, but just as I was about to do so I saw a photograph of my son inside and then a folded scrap of paper with the words "Don't open before 2100. This note is from 2002, Sweden."

Of course, I opened the note and read these words, written in pencil:

Hello, this is a time capsule! Have saved things that meant a lot to me. The photo is of me, Gunnar, aged 10. The paper is a membership card for Hammarby Football Club. The metal capsule is from 1999 and might be worth a lot when you find it. The flat metal thing is a Swedish 1 krona coin from 1973.

That was it.

I sat down on the floor and remembered how I had done the same thing when I was my son's age, burying treasures, squeezing notes into little crevices in the house where I grew up. And I did so with the innate, completely natural fascination that any 10-year-old has for the dizzying eternal perspectives of time. What came before me? What will come after me? What will my contribution be to the world?

Sitting there, I wonder when we grown-ups lose this perspective. How could we regain it? And what will we leave behind to the world of 2100?

WHAT CAN WE EAT?

September 2011

In the August 2011 issue of *Current Biology*, a very sad article appeared. Researchers from the Clemson University in South Carolina, USA, had checked, using DNA markers, to what extent fish labelling in the grocery stores was correct and found that almost one fish out of ten was mislabelled, not only as coming from the wrong geographical region, but according to population biologist Peter Marko, as being "actually an entirely other species".

But the saddest thing about this isn't only that fish consumers are fooled into buying fish that may be illegally or unsustainably caught – it is that this was also true for fish certified by the Marine Stewardship Council (MSC). Of the Chilean sea bass (brand name for Patagonian tooth fish) certified as sustainable by this almost world-encompassing organisation, some 15 per cent were genetically distinct from fish previously collected from the certified fishery; one sample even came from the other side of the globe.

The same summer of 2011 another sad news story emerged in the British news, when researchers from Dublin University released similar findings. In the UK 7 out of 96 samples of fish were mislabelled, in Ireland 31 out of 131 samples – 28 per cent! – were mislabeled! And of course the researchers did not only go into one shop – the samples were collected from six different large retail chains that source their products from nine suppliers.

More than half of the mislabelling allowed Atlantic cod from depleted stocks to be sold as sustainably sourced Pacific cod, and the researchers did not believe one minute that this was done by accident, rather that it was a deliberate fraud to meet the demand of the UK market for sustainable seafood. And even worse, the researchers suggested to the UK website fish2fork.com:

"When quota limits are reached, IUU (Illegal, Unreported and Unregulated) catches of this species could be disguised when these products are mislabelled as foreign imports."

Both these studies of course raise great concerns regarding the validity of the chain of custody documentation both of the big food retail outlets, and also of the world's largest sustainable seafood certification scheme, the MSC. And both cases prove the need for stricter legislation and control of the world's fisheries, including establishing a worldwide register of fish DNA, which would make it a simple process for fish importers, retailers, or NGOs to check whether or not the fish is what it is claimed to be.

But to create the political will to act there seems to be a need to assure politicians that this is what people want, actually what the consumers *demand* – and what better way to show this, than by using your wallet as the instrument of power it really is? In Sweden sales of cod went down some 50 per cent after the Swedish edition of this book was published, and all the major retailers actually dramatically changed their fish policies. One of the biggest ones, Coop, went as far as doing a complete audit of its sea-footprint, including the use of "shadow-fish" in aquaculture products, pig and hen production, and dog and cat food, as well as checking the source of all the fresh fish sold during the summer times by local fish-dealers outside the stores. Now they know exactly what fishing methods are used, the amount of fuel used, CO_2 emissions per kilo of fish, and how large an area is being bottom trawled each year in the world's oceans just to supply Swedish Coop. They are also active in the European retailers' organisation Eurocommerce in spreading this method of fish auditing, and they are also trying to set up common goals to reduce all the negative impacts on the sea environment from European consumers.

Another major impact of the consumer pressure in Sweden has been that the state-owned Norwegian fish export organisation became extremely worried about the future of their cod exports to Sweden, and as a result most of the Norwegian cod fisheries in the Barents Sea are now environmentally certified.

But if Swedish consumers hadn't started asking for sustainable seafood in the stores – what would have happened then? Probably not much. So – even if consumer power isn't the perfect instrument, and we as consumers can still be fooled by dishonest fish producers or too lax controls by those who are supposed to guarantee that their certification makes a certain fish product safer than another – we still can't miss the chance to do what we can. Three or four tricky questions a day at the fishmonger's, or at the grocery store – and the people selling fish will certainly feel the pressure to try and get better answers from their suppliers. And if consumers actually boycott fish – then we would see changes sooner than we ever would have believed possible! Remember that the news of suspected bacteria in Spanish cucumbers actually wrecked a whole industry in just a few weeks. Consumers are powerful; never believe anything else!

So – how do we start? Really hoping I am not discouraging those of you ready for action too much, I just have to emphasise again that buying sustainable fish isn't at all an easy thing to do. Fish belonging to the same species can be sustainably caught in one part of ocean, and not in another. The methods used in the fishery must also be taken into consideration – so even if the stock of grey shrimps in the North Sea for instance is extremely abundant – the trawling method of catching them is normally not sustainable – it can actually cause a by-catch of up to 80 per cent, which is totally absurd. Furthermore, the shrimps are then transported to Morocco to be hand peeled, and then brought back again to Europe, of course drowned in not so appetising amounts of preservatives. And the CO_2 emissions of the transportation cannot be regarded as environmentally friendly.

Methods, locations and human health aspects aside, we must also consider the size of the fish. It may be that you consider buying fish from a species or a stock that is slightly overfished – but then you at least don't want to buy a juvenile of that species – so sometimes you have to know what a "normal" size fish is, and perhaps not buy a fillet of Baltic cod smaller than 30 centimetres.

Farmed fish has of course its downsides, as explained in chapter nine, but some is better than others and should be judged more case by case. The certifying body KRAV has now given the green light to salmon producers in Norway, which gives hope of improved conditions for farmed salmon and the use of shadow-fish.

Having said all this, I refrain from making a list in this book of species that are sustainably fished, since in general terms it would be really short (in Europe it would contain only three species: saithe, sprat and mackerel, and soon mackerel might have to be de-listed because of vast Icelandic and Faroese IUU fishing). Instead, I would recommend readers to go to the World Wildlife Fund's website www.wwf.panda. org to find more detailed lists of sustainable seafood, where distinctions are made between farmed fish and wild-caught, different stocks, catch methods and labels. WWF have collected seafood guides from a large number of different countries, in very many different languages, from Finnish and Polish to Indonesian and Catalan.

Specifically for the US there is of course also the Seafood guide from the Monterey Aquarium possible to download for free at the App-store, and surely more seafood guides for smart phones will appear soon. Until then all of the above are possible to print in small formats so that you can carry them in your wallet.

Finally, I wish I could stop here by saying that the safest way is always to choose the certified products in the stores, but – although I strongly support the voluntary efforts made by all the fishermen in the world to meet the criteria put up by these certifying bodies – we, the consumers, should never just accept that this is the best way of fishing. MSC and other certifying bodies should always strive to improve, always strive to be at the forefront, revising their criteria when they become "too easy" to fulfill. "Legal" should be fundamentally sustainable in my view, and environmentally certified should be top class animal friendly, climate friendly and a hundred per cent transparent when it comes to traceability.

So – lots more to do!

Let's do it – together.

REFERENCES AND SOURCES

CHAPTER 1: THE EEL

State of the European eel stock: International Council of the Exploration of the Seas **(ICES):** *Eel stock dangerously close to collapse,* http://www.ices.dk/marineworld/eel.asp(Willem Dekker, Netherlands Institute for Fisheries Research).

State of European eel stock, ICES latest advice 2011: http://www.ices.dk/committe/acom/comwork/report/2010/2010/eel-eur.pdf

Decrease of Swedish eel catches since the 1960s: Finfo 2005:3. Niklas B. Sjöberg and Erik Petersson: *Blankålsmärkning. Till hjälp för att förstå blankålens migration i Östersjön,* p. 14.

The EU Commission's proposal for an eel recovery plan: *Establishing measures for the recovery of the stock of European Eel,* Council regulation, 2005.

Eel life cycle: Marianne Köie and Ulf Svedberg: *Havets djur,* (Prisma, 1999).

Restocked eels' failure to find their way out of the Baltic Sea: Lars Westin: *Migration failure in stocked eels Anguilla anguilla,* (Marine Ecology Progress Series Vol. 254:307-311. 2003).

Restocked eels potential to migrate: Limburg et al: "Do Stocked Freshwater Eels migrate? Evidence from the Baltic Suggest 'yes'"(American Fisheries Society Symposium 33:275–284, 2003).

Value and weight of landed eels in the sea and in lakes in Sweden 2006: Statistics Sweden, Statistiska meddelanden JO 50 SM 0701: "Saltsjöfiskets fångster under december 2006 och hela 2006" , and Statistiska meddelanden JO56 SM 0701: "Det yrkesmässiga fisket i sötvatten 2006".

CHAPTER 2: THE ALARM BELLS TOLL

Greenpaper on the future of the Common Fisheries Policy: Com/2001/0135 final (2001-03-20).

Board of Fisheries impact assessment of a unilateral Swedish cod fishing moratorium: Dnr: 43-2362-02.

The closure of parts of Georges Bank: Carl Safina *Song for the Blue Ocean* (Owl Books, 1997) p. 44.

Consumption of diesel during trawling for nephrops: "Branschen vill begränsa miljövänligt kräftfiske" Swedish daily newspaper *Dagens Nyheter* (2007-02-21).

Discarding of fish: Board of Fisheries "Angeläget att minska dumpning" *Sött&Salt* (issue 1:2004).

Evaluation of the effect of EU intervention prices on the fish market: Board of Fisheries "Utvärdering av återtagssystemet" Dnr 121-2351-00.

Economic income of professional fishermen: Board of Fisheries report 2000:1, "Kustfiskebefolkningens ekonomi".

Value added and contribution to GDP: Statistics Sweden, National Accounts.

Turnover of the horse industry: Ministry of Agriculture, Regeringens skrivelse 2003/04:54.

Value of moose hunting: oral information Bo Toresson, former chair of Swedish hunters' association.

Value of cod catches during the last century: *Svenska fiskets framtid och samhällsnytta,* KSLA-tidskrift 10/2001, p. 19.

Value of cod catches: Statistics Sweden, Statistiska medelanden JO 50 SM 0701.

CHAPTER 3: THE SILENCE OF THE ALARMS

The coelacanth: Samantha Weinberg: *A fish caught in time: the Search for the Coelacanth* (Perennial, 2001).

Coral reefs in Norway: Richard Ellis: *The Empty Ocean* (Island Press) p. 276.

Jacques Piccard's and Don Walsh's descent into the Mariana Trench: Sylvia A. Earle: *Sea Change – A message of the Oceans,* (Fawcett Books, 1995) p. 48.

Open letter from the Swedish Anglers Association concerning the no-trawl zone: http://www.sportfiskarna.se/aktuellt/pressm. asp?Id=128.

Director-General Karl-Olov Öster's answer to the open letter: http://www.skargardsbryggan.com/dokument/svar_sportfisk_brev. pdf.

The status of coastal cod by the Swedish west coast and in the Öresund strait: Henrik Svedäng, Vidar Öresland, Massimilliano Cardinale, Hans Hallbäck, Peter Jakobsson: *De kustnära fiskbeståndens utveckling och nuvarande status vid svenska västkusten, Torskprojektet steg I-III,* Swedish Board of Fisheries Marine Laboratory of Lysekil, 2002.

Visual sense of fish: Leif Andersson, Björn Röhsman "Under vattenytan" (Spektras handboksserie, 1983).

CHAPTER 4: THE TRAGEDY OF THE COMMONS

"The tragedy of the Commons": Garret Hardin, *Science* 162, 1243-1248, 1968. http://www.sciencemag.org/sciext/sotp/commons.dtl.

Overfishing of highly migratory species such as tuna and swordfish: Stephen Sloan: *Ocean Bankruptcy – World Fisheries on the Brink of Disaster* (Lyons Press, 2003).

History of cod fishing: Mark Kurlansky: *Cod – A Biography of the Fish That Changed the* World (Penguin, 1998).

Magnitude of sports fishing: Board of Fisheries and Statistics Sweden: *Fiske 2005 – En undersökning om svenskars fritidsfiske.* (Finfo 2005:10).

Recreational fishermen's "willingness to pay": TemaNord 2000:604 (Nordiska Ministerrådet).

Recreational value of sports fishing: *Sportfiskets betydelse och samhällsnytta*, report by Ingemar Norling, the section for Health Care research, Sahlgrenska University Hospital, Göteborg, 2003.

Cost of adipose fin clipping: "Fiskevård", 1/04 (Swedish Water Owners Organisation).

First nation people's view of salmon: Carl Safina "Song for the Blue Ocean" p 140 (Owl Books, 1997).

Obligation to anaesthetize farmed salmon: Swedish Animal Protection Law (1988:534), Djurskyddsförordningen (1988:539) and Statens jordbruksverks föreskrifter (SJVFS 1993:154).

The double roles of the Board of Fisheries; both protecting and controlling the fishing industry: Johanna Eriksson: *Så bereds en torsk inför behandlingsbordet, från vetenskap till politik* (Institutionen för tematisk utbildning och forskning, Programme of Environmental Studies, University of Linköping, 2002) p. 15.

The economic cost of fisheries management, the "Fish and Fraud" report: ESO Ds 1997:81: *Fisk och Fusk, Mål, medel och makt i fiskeripolitiken*, p. 20.

CHAPTER 5: BEFORE THE BIG BOATS CAME

History of Smögen: Casper Ljungdahl: *Smögen och Hasselösund – bohuslänska fiskesamhällen i förändring – människorna på Sotenäset och vid Skagerrakkusten* (Munkedal 1993).

The competition with the Göteborg boats: Sportfiskarna i Väst, och Kommittén Värna Västerhavet: Värna *Västerhavet! Medan tid är...* (Bokförlaget Settern 1994).

Reply to the ministry report on sustainable development in the coastal communities: Norra Bohusläns Producentorganisation: *Yttrande över Miljövårdsberedningens betänkande SOU 1996:153 "Hållbar utveckling i Sveriges skärgårdsområden".*

EU market and structural funds granted to Smögen: Board of Fisheries, Marknads- och strukturavdelningen: *Investeringskatalog Områden utanför mål 1 – förteckning över investeringar som beviljats strukturstöd i form av avskrivningslån inom fiskerisektorn 2000-2006.*

The fate of spiny dogfish in North America: Richard Ellis *The Empty Ocean* (Island Press 2003), p. 48.

Behaviour of spiny dogfish, as witnessed by sports fishermen: http://havsfiske.wasa.net/info/Pigghaj.html.

New quota for spiny dogfish: Sött & Salt, newsletter from Board of Fisheries *Kvoter införs på hotad pigghaj* (2007-02-26).

Catch history of spiny dogfish: Board of Fisheries *Fiskbestånd och miljö i hav och sötvatten, resurs- och miljööversikt 2006*, p. 45.

Lack of political guidance of the fisheries management: Nils Gunnar Billinger: ESO: *Är regeringskansliet för stort, för litet eller lagom?*, seminar, government office of Rosenbad, 1998.

Fisheries policy as "the small minority's hostage": Miljövårdsberedningens promemoria *Strategi för ett hållbart fiske* (2006:1), p. 21.

CHAPTER 6: THE SUBSIDIES

On the Atlantic Fisheries Adjustment Package: Charles Clover: *The end of the line – how overfishing is changing the world and what we eat* (Ebury press 2004), s 116.

Decrease in effectiveness in Swedish fisheries since Sweden's membership of the EU: Joacim Johannesson and Tore Gustavsson: *Fuelling fishing fleet inefficiency*, report from Board of Fisheries 2005-06-30.

Cost of Board of Fisheries: Fiskeriverkets årsredovisning 2006.

Payments of fishermen's unemployment benefits: http://www.handels.se/akassan/yrkesfiskare/.

Cost of fisheries control: *Årsredovisning 2006*, www.kustbevakningen.se.

The effect of controls on catches: "Den svenska fiskerikontrollen", SOU 2005:27.

Fuel consumption in the fishing sector: *Energianvändning inom fiskesektorn 2005*, Statistics Sweden, SCB 2006.

Evaluation of the effect of EU intervention prices on the fish market: Board of Fisheries "Utvärdering av återtagssystemet" Dnr 121-2351-00.

Report on situation for professional fishermen, disputed by experts due to proposal of a special tax break: Ulf Lönnqvist: "Yrkesfiskets konkurrenssituation" SOU 1999:3.

Proposal for a special tax break for professional fishermen by MP Eskil Erlandsson, later to become agricultural minister: "Småskaligt yrkesfiske" 1999/2000:MJ406.

External environmental costs of the maritime sector: Per Kågesson: *Internalisering av sjöfartens externa kostnader* (2000-11-27).

CHAPTER 7: THE FISHERIES AGREEMENTS WITH THE THIRD WORLD

The "advantageous" protocol with Cape Verde: European Parliament draft report 2004/0058 http://www.europarl.europa. eu/RegData/commissions/pech/projet_rapport/2004/346863/ PECH_PR(2004)346863_EN.pdf.

Market price of bluefin tuna: Stephen Sloan *Ocean Bankruptcy – World Fisheries on the Brink of Disaster* p. 102 (The Lyons Press 2003).

Fish consumption in the world: FAO. *The State of World Fisheries and Aquaculture 2006*, pp. 36-37.

The Senegal agreement and the EU: Charles Clover *The end of the line – how overfishing is changing the world and what we eat* (Ebury press 2004), p. 37.

The case of Namibia: Staffan Danielsson: *Hur den fattiges fisk hamnar på den rikes bord* (Globala studier nr 14 2002), p. 18.

The EU advantages of fisheries agreements: Mikael Cullberg *EU:s fiskeriavtal med utvecklingsländer – från resurstillträde till partnerskap?* Finfo 2005:2, p. 59.

Effects of fisheries agreements in developing countries: Susanna Hughes *Effekter av EU:s avtal om fiske i u-länder*, Livsmedelsekonomiska institutet 2004:6.

Less fish leads to more demand for bushmeat: Ola Säll *Afrikas aptit på vilt växer*, Svenska Dagbladet, 2006-06-18, and Karin Bojs: *EU:s överfiske dödar elefanter och apor*, Dagens Nyheter 2004-11-13.

CHAPTER 8: THE EU AND THE DUTY TO EXPLOIT

Combined value of world's fisheries subsidies: Charles Clover *The End of the Line*, p. 136.

Critique against EU shark fishing policies: http://www.sharkalliance.org.

EU Commission, Directorate-General for Fisheries and Maritime affairs: http://ec.europa.eu/dgs/maritimeaffairs_fisheries/index_en.htm.

Committee on Fisheries in the European Parliament: http://www.europarl.europa.eu/activities/committees/homeCom.do?body=PECH.

The row concerning Steffen Smidt and the reform of the Common Fisheries Policy: Tom Hansson: *Oortodoxa metoder för att stoppa EU:s fiskerireform*, Dagens Forskning Nr 10, 2002-05-13/14, *Stormen runt fiskeripolitiken fortsätter*, Dagens Forskning Nr 11, 2002-06-10, *Europas fiskeflotta skärs ner*, Dagens Forskning Nr 12 2002-06-10, http://www.acc.umu.se/~widmark/eu-fiske.html.

CHAPTER 9: AQUACULTURE – THE BEST SOLUTION?

Increase in Norwegian fishing capacity: WWF: *The Barents Sea Cod – The last of the large cod stocks,* May 2004.

Emissions of nutrients from aquaculture plants: Nils Kautsky, Carl Folke, Max Troell och Patrik Rönnbäck. *Torskar torsken? Forskare och fiskare om fisk och fiske* (Formas, 2003), p. 99.

The use of wild fish in aquaculture: Swedish Environmental Agency: *Förändringar under ytan,* and *Torskar torsken? Forskare och fiskare om fisk och fiske* (Formas, 2003), p. 96.

The view of Norwegian producers of farmed salmon: Eksportutvalget for fisk, http://www.godfisk.no.

Decline and threat to coastal Norwegian cod: Havforskninginstituttet: *Havets resurser og miljö 2005,* p. 69, ICES stock assessments: http://www.ices.dk/committe/acom/comwork/report/2011/2011/cod-coas.pdf.

Diseases in farmed Norwegian fish: Havforskningsinstituttet: Kyst og havbruk, 2005.

Francisella sp. in Swedish cod: Ny torsksjukdom drabbar vilda bestånd, SVA-vet 2/2006.

CHAPTER 10: THE WAY FORWARD

Results of eel tagging : Håkan Westerberg: *Resultat av ålmärkning i Östersjön*, Swedish fishermen's trade journal, Yrkesfiskaren nr 23/24-06.

Composition of RAC: http://ec.europa.eu/fisheries/press_corner/press_releases/archives/com04/com04_23_en.htm.

The North Atlantic in the 1500 and 1600s: Daniel Pauly and Jay Maclean: *In a perfect Ocean – the State of Fisheries and Ecosystems in the North Atlantic Ocean* (Island press 2003), pages 8–9.

"Shifting baselines": Daniel Pauly and Jay Maclean: *In a perfect Ocean,* p. 29, or www.shiftingbaselines.org.

Overfishing of oysters in Chesapeake Bay: Jack Sobel and Craig Dahlgren: *Marine Reserves – A Guide to Science, Design and use* (Island Press 2004), p. 36. And Daniel Pauly *In a perfect Ocean* pages 19–20.

The share of discarded fish: *Strategi för ett hållbart fiske,* Miljövårdsberedningens promemoria 2006:1, p. 39.

The history of trawling: Mark Kurlansky: *Cod – A Biography of the Fish That Changed the World* (Penguin, 1998).

Visual effects of bottom trawling: http://www.ccb.se/documents/Bottentralningsve.pdf.

State of the sea outside Cape Canaveral: *Marine protected areas,* Los Angeles Times (2002-07-22).

On marine reserves: Rod Fujita: *Heal the Ocean – Solutions for Saving our Seas* (New Society Publishers, 2003); Jack Sobel and Craig Dahlgren: *Marine Reserves – A Guide to Science, Design and use* (Island Press 2004).

Looking for suitable no-take marine areas in Sweden: Board of Fisheries: *Inrättandet av ett fiskefritt område* (2006-02-27).

Goat Island Marine Reserve: Charles Clover *The End of the line,* (Ebury press, 2004) p. 215.

Fishing down the foodweb Daniel Pauly, *In a Perfect Ocean,* pages 53-56.

Acoustic transmitters on cod: James Lindholm, Pfleger Institute of Environmental Research, www.pier.org.

Increase in cod on Vinga artificial reef: Board of Fisheries: *Effekter av fredningsområden på fisk och kräftdjur i svenska vatten*, (2006-02-21), p. 11.

The cod in the Öresund strait, the effects of protecting fish in Swedish waters: Board of Fisheries: *Effekter av fredningsområden på fisk och kräftdjur i svenska vatten* (2006-02-21), p. 17.

The choice of the island of Gotska Sandön, and the end of cod and turbot fisheries: Board of Fisheries: *Inrättande av ett fiskefritt område*, (2006-02-27), pages 3 and 8.

Illegal trawling in Öresund; *Framtid för fiske i Öresund* – report from seminar held in Helsingborg and Helsingör 17-18 of November 2003, and Peter Nilsson: *Fusket som tömmer haven*, Sveriges Natur nr 6/2001.

The halibut race in Alaska: Charles Clover: *The End of the Line,* p. 211, Rod Fujita: *Heal the Ocean*, p. 109.

Discarding of smaller fish, so-called *high grading*: Daniel Pauly: *In a perfect Ocean*, p. 16.

ITQs: Board of Fisheries: Sött & Salt, *Kvoter, zoner och andra sätt att fördela fisket 2/2004. Fiskeriverket vill ha effortreglering, samförvaltning & individuella kvoter* (2005-10-03).

How Australia and New Zeeland finance their fisheries management: Daniel Pauly: *In a perfect Ocean*, p. 137.

The hunt for the pirate ship Viarsa: G Bruce Knecht: *Hooked – Poaching Pirates and the Perfect Fish* (Rodale, 2006).

The extent of unreported Swedish catches: Report Dnr 121-3160-02, p. 12. Board of Fisheries, section for control, 2004-06-01. Håkan Eggers and Anders Ellegård: *Fishery control and regulation compliance: a case for co-management in Swedish commercial fisheries,* Marine Policy 27 (2003), pages 525-533, Board of Fisheries, Sött & Salt *Fiskeriverket bekräftar orapporterat fiske*, (2007-03-15).

Punishment in Sweden for fishing illegaly: Board of Fisheries: *Den svenska användningen av sanktioner i ärenden om överträdelser i fiske-bestämmelser – erfarenheter och förslag.*

CHAPTER 11: TWO SECONDS

Sylvia A Earle: *Sea Change – A message of the Oceans*, (Fawcett books, 1995).

Bengt Hubendick: *Mot en ljusnande framtid*, (Gidlunds, 1991).

Theodore Roszak: *The Voice of the Earth*, (Phanes Press 1992 and 2002).

James Lovelock: *The revenge of Gaia – Earth's Climate Crisis and the Fate of Humanity*, (Basic Books, 2006).

Rod Fujita: *Heal the Ocean – Solutions for Saving our Seas*, (New Society Publishers, 2003).

Linda K Glover, Sylvia A Earle: *Defying Oceans End – An Agenda for Action*, (Island Press, 2004).

Rachel Carson: *The Sea Around Us* (Oxford University Press, 1950, 2003).

THANKS

to all the people who helped me with the Swedish edition; almost all of whom are mentioned in this book. My special thanks to Henrik Svedäng, Michael Earle and Dr Daniel Pauly who checked the book and gave invaluable advice and comments, the Foundation Baltic Sea 2020 who helped with the English translation, and my family who have endured this long and sometimes tiresome process of creating this book, from writing to publication.

Last but not least, my thanks to the Swedish Greens who believed in me, and to all the Swedish people who voted for me after having read this book. Knowing that I represent all of you makes it a lot easier for me to continue to struggle inside the system in the European parliament.

I also thank you all for keeping up the fight – hoping now that consciousness and the will for change will spread all over Europe.

CPSIA information can be obtained at www.ICGtesting.com
Printed in the USA
BVOW03s0410190515

400748BV00014B/272/P

9 781908 341532